STRING, STRAIGHTEDGE, AND SHADOW
THE STORY OF GEOMETRY

ILLUSTRATED BY CORYDON BELL

JULIA E. DIGGINS

STRING, STRAIGHTEDGE, AND SHADOW

THE STORY OF GEOMETRY

JAMIE YORK PRESS

To Maggie, Peter, Marijo, and May,

and all the children whom I have taught

whose interest, enthusiam, and inquiring minds

made me want to write this book.

Copyright © 1965 by Julia E. Diggins
All right reserved
First published in 1965 by The Viking Press, Inc.

This edition published in 2012 by

JAMIE YORK
PRESS

*Meaningful Math Books for Waldorf
Public, Private, and Home Schools*
www.JamieYorkPress.com

ISBN: 978-1-938210-09-9

CONTENTS

PROLOGUE: THREE WONDERFUL TOOLS 7

THE BEGINNINGS: GEOMETRY AND NATURE 11
    1. A Sixth Sense 12 • 2. From the Art Gallery of the Universe 15 • 4. Secrets of the Stone Age 21

IN THE ANCIENT NEAR EAST: GEOMETRY AND DAILY LIFE 28
    4. Reading the Shadows 29 • 5. The Rope-Stretchers 36 • 6. The Star-Gazers 47

THE IONIAN GREEKS: GEOMETRY AND THOUGHT 58
    7. The 6th Century B.C. 59 • 8. Thales at Home and Abroad 63 • 9. How High is the Pyramid? 69 • 10. The Rules of the Game 79

THE SECRET BROTHERHOOD: GEOMETRY, MATHEMATICS, AND MAGIC 92
    11. Pythagoras and His Followers 93 • 12. A Famous Theorem • 13. "Dice of the Gods" 106 • 14. The Unspeakable Tragedy 115

FROM THE ACADEMY TO THE MUSEUM: GEOMETRY, ART, AND SCIENCE 123
    15. The Golden Age and The Golden Mean 124 • 16. A Royal Road, After All 139 • 17. The Whole, Round Earth! 149

INDEX 156

MORE TITLES BY JAMIE YORK PRESS 159

But by the beauty of shape I want you to understand not what the multitude generally means by this expression, like the beauty of living beings or of paintings representing them, but something alternatively rectilinear and circular, and the surfaces and solids which one can produce from the rectilinear and circular with compass, set square, and rule. For these things are not like the others, conditionally beautiful, but are beautiful in themselves.

—Plato

PROLOGUE

Three Wonderful Tools

The *string*, the *straightedge*, and the *shadow*—they are easy to find almost anywhere. A string can usually be found in a boy's pocket; a straightedge in a desk drawer. Shadows are constant companions on a sunny day.

Yet these are also three wonderful tools. Using them alone, ancient men discovered the ideas and constructions of elementary geometry more than twenty centuries ago. It was all done with just string, straightedge, and shadows! And that is the subject of this story.

Nowadays, men build bridges that span the Golden Gate and the Mackinac Straits. They make planes that go six times faster than the speed of sound, submarines that circle the globe without surfacing, missiles that reach outer space. They drill oil wells deeper than the height of Mount Everest. They have harnessed the power of the tiny atom, and placed men in orbit around the earth. But behind these 20th-century wonders lies a long history packed with beauty, adventure, and struggle.

Through the ages, men have searched to find the secrets of the universe. As these secrets were learned, they were written down in mathematical symbols. Today the search for secrets goes on: the mystery of the universe is still unfolding. Even the huge telescope at Mount Palomar is only a tiny window into a vast unknown, and when we finally travel to the moon and planets, there will be other worlds circling other stars for us to visit. In this unending search, mathematics has been a key from the very start.

A long, long time ago primitive men observed the lines and curves and other forms of nature. They marveled at them, and copied them as best they could.

Then, as need arose at the dawn of history, men learned to construct these lines, curves, and forms accurately. They used the *string* to trace a circle, to lay off a right angle, to stretch a straight line. They used as a *straightedge* anything else with which they could draw a straight line. They came to realize that *shadows* are the sun's handwriting upon the earth to tell the secrets of order in the universe.

By using these easy-to-find tools, the early civilizations learned how to tell time and direction. They designed their dwellings, temples, and tombs, laid out their fields, and constructed irrigation ditches. They started to measure and record the apparent travels of the sun, moon, and stars. They found ways of guiding their own travels across the sea and the trackless plains.

So, during thousands of years, the ancient peoples of the Near East built up a practical art that came to be known as geometry —from *geo*, "earth," and *metria*, "measurement."

And then other peoples—the Greeks—turned from the tools to the rules.

Thinking men began to question and wonder *why*. They were still practical people, but they were also interested in abstract rules about lines and curves and angles.

For centuries, many men worked on these rules. Some studied beautiful geometric forms and tried to combine them with numbers into a pattern of the universe. One solved useful mechanical problems, such as how to raise water out of the hold of a leaky vessel. Others worked on "useless" puzzles, real brain-twisters.

Out of their work came *theoretical geometry*.

The Greek geometers developed logical ways of thought. They discovered that the earth is a sphere, and they measured the girth of this sphere and the tilt of its axis. They discovered the properties of curves which they called the *ellipse*, the *parabola*, the *hyperbola*, and the *spiral*—curves that many centuries later were recognized as the paths of motion for bodies in space. They helped lay the foundations of modern science and invention.

But all that took an unbelievably long time and happened very, very slowly. Nowadays we live in an age of speed: new models of automobiles and television sets and sound-reproduction systems are perfected before the old ones are worn out or even paid for. So it may be hard to realize just how long it took to make those ancient discoveries.

String, Straightedge, and Shadow

For thousands of years, billions of men have lived on the earth, and only a limited number of them have contributed to the development of geometry. Many of the "facts" are missing from the early part of their story. For the most faraway period, the records are gone, and sometimes we have only a name or a legend.

So in this book, we shall try to tell a *story* instead—only the high lights, the best characters, the dramatic incidents.

This will be the thrilling story of geometry in the ancient world, from the earliest prehistoric men to the best-selling text in the history of mathematics, the *Elements* of Euclid. The story really is a thriller, with wonder and adventure and magic, and even a murder mystery!

As you read it, you will see that it is a timeless tale besides. For the old discoveries are very much alive today. And all of them were made with just three wonderful tools: the string, the straightedge, and the shadow.

THE BEGINNINGS

Geometry, and Nature

1. A SIXTH SENSE

Geometry itself—if we trace its deepest roots—goes back long before the discovery of its tools. It goes back even before the first observations of primitive man. In fact it goes way back to nature and life and a "sixth sense" that is in every one of us.

This mysterious faculty can be called an inborn sense of mathematics.

Because we are all part of this immense universe, and are bound by its laws, we have a natural sensitivity to its order and beauty. All men are *thinking* parts of the universe, so they have used their sensitivity to translate these laws of order and beauty into mathematical language.

You can understand this sensitivity from your own personal experience. When you shook a rattle and recognized rhythm, or rolled a ball in your playpen and noticed the characteristics of a geometric form, your study of mathematics began.

If you ever have entered a room and adjusted the curtains or straightened a picture on the wall, you have used an innate sense of measurement. If you have stopped your ears to shut out strident or confusing sounds, you have sensed a desire for harmony.

When you first realized security in the fact that daylight and darkness follow a constant pattern, you experienced a sense of order in the universe.

All living things proclaim this order and have a sense of mathematics that comes naturally.

Birds, bees, whales, and seals have a natural sense of direction and distance. Why does a flock of birds fly in angular formation and a family of ducks swim along the creek in the same angular

form? How do birds fly back and forth with the seasons, finding their way to the same resting place? How do bees direct one another to the source of honey? They never studied the science of navigation—yet they seem to know how to navigate.

In their building, too, the creatures show a sense of form. Bees construct their six-sided cells in the most efficient method of space-filling. But they never studied architecture. Spiders spin an almost perfect spiral web, yet they never studied engineering. Few birds fail to observe the principles of symmetry in the structure of a nest. And all animals seem to know that a straight line is the shortest distance between two points. It is almost as though they had instruments built into them.

Within ourselves also there seem to be built-in measuring instruments. A natural compass helps our sense of direction. A sense of mass and weight keeps us from trying to lift too heavy an object. A sense of symmetry guides us in hanging a picture, or spreading a tablecloth, or adjusting a bedspread. A natural sense of rhythm makes us tap a foot or want to dance when we hear music.

And some of us have developed this special sense in our work.

Artists, for instance, are natural mathematicians. For the secret of beauty is order. Artists must continually compare sizes and distances in their relation to one another in the composition of a picture. They have to put form and atmosphere on a flat rectangular canvas. So they have to judge the values of tones from light to dark, and the force of contrast, and the intensity of color.

Musicians use intuitive mathematics, too. They must keep time, or have a conductor do it for them. They must know the value of notes and rests within a measure. They have to judge the loudness or softness of a tone, and how long to sustain a note. With their fingers, pianists have to measure the speed and strength of their touch.

And what about dancers? The whole plan and execution of a ballet is based upon an amazing precision of time, motion, and pattern, all combined with the rhythm of the orchestra.

Poets, too, have to measure the time-beats in a line, and the related beats from line to line. They have to judge words not only for meaning and rhyme, but for the number of syllables and the strength of the accents inside the words.

Some people, as you see, have developed their "sixth sense." But within all of us there is a natural mathematical sense tuned to the natural order of the universe. We all like order and harmony. We like things in proportion; when they seem to be out of proportion we try to correct them.

It was from this inner sense—man's sensitivity to the order and harmony of the universe—that geometry really began.

2. FROM THE ART GALLERY OF THE UNIVERSE

When and how did geometry begin? Who first discovered the lines, curves, and shapes that we call simple geometric forms?

These forms were discovered by the earliest men who wandered on the planet earth. For they were forms found everywhere in nature—in the vast art gallery of the universe.

Let us return in our imaginations through tens of thousands of years, to the time when the first men wandered alone or in small groups over the earth. Future masters of the earth, they were still cringing in fear. All the great secrets and wonderful resources were locked away from them, awaiting the key of discovery.

They hid from the lightning. They were severely frightened by the seemingly blind and ruthless forces of nature. They thought that, as the days grew shorter and the sun sank lower, daylight would disappear forever and leave them in the chilly darkness.

So they huddled by their precious fires.

Fire—that was the first great secret wrested from nature. Prehistoric men tended the fires set when lightning struck trees, and learned to make their own fires. But this did not dispel their dread that the sun was dying. All primitive peoples have shared the fear, and rites and chants and sacrifices were developed to help the sun's return.

Gradually, however, the returning warmth and light raised their spirits. This cycle had to occur again and again through centuries before it gave them confidence in its dependable pattern.

But slowly they began to absorb a feeling of rhythm and harmony and order in nature. Terror gave way to wonder and fresh discovery.

Then they began to listen to the music of the wind, the beat of the rain. On warm nights they heard the insects' song and the symphony of the frogs. They noticed the time-beat of their hearts and the rhythm of breathing.

They began to observe lines. The jagged lightning was terrifying. But they felt the tranquil peace in the line of the far horizon, like the position of their own bodies when they lay down to sleep. They admired the strong permanence in the line of a tall, straight tree, like their own positions when they stood up. They saw the sadness in the curve of a wilted stem, and the airy lightness in the same curve on a soaring cloud.

To the early men who trudged in fear through the wonderful art gallery of the universe, everything was a new discovery and an amazing surprise.

The Beginnings

Before we learn how geometry came out of these discoveries, you may want to tour the same art gallery for yourself. Just look around and you will find the circle, right angle, triangle, square, five-sided pentagon, six-sided hexagon, and spiral, beautifully revealed by nature in sky, earth, and sea.

Gaze at the immensity of the night sky! The greatest natural law that we all feel and appreciate is that of order over chaos: it can be seen in the patterns of the starry firmament—not just those visible to the naked eye, but the ones disclosed by telescopes and calculations. In outer space, distant galaxies—immense clusters of stars—unfold in gigantic spirals. Closer to us, planets and comets swing around the sun in elliptical orbits. Meteors blaze into our own atmosphere in parabolic curves.

Or look through a microscope at the intricate beauty of crystals! For centuries down under the earth, under its great heat and pressure, minerals have been solidifying into crystalline form. Crystals are the flowers of the mineral world: they "grow" by building onto themselves the same materials that they are made of. One mineral can be told from another by its crystal, for that is the geometric form a chemical substance assumes when changing freely from liquid or gas to a solid. Large or small, regular or irregular, all crystals of a given mineral have the same inner lattice structure and the same relations between faces and axes.

Have you seen quartz crystals? They are six-sided (or hexagonal) prisms capped by a six-sided pyramid. If you pound one to powder and put the powder in a solution to recrystallize, the resulting crystal takes the same shape: a hexagonal prism capped by a hexagonal pyramid!

Why? Because nature always tends toward simple geometrical

shapes. These are the result of universal laws of inner structure and outer symmetry: the same forces that mold a teardrop mold a star.

And these forces are at work everywhere in the geometry of living things.

Did you ever wander through forests and fields in the early spring? The woodland bursts into bloom with the tiny triangles of the fresh little three-petaled trillium, the square form of the chalk-white, four-petaled dogwood, and the pentagonal blossom of the mountain laurel. The fruit trees swell out with five-petaled flowers. Bend close to the ground and see the tiny fern opening, topped with delicate spirals, and notice the spiraling tendrils on a vine.

The Beginnings

Stroll across a meadow covered with five-petaled buttercups and the circular faces of daisies and dandelions. And when you pull up a dandelion, stop and consider its beauty, from the spiral growth of the plant to the gossamer sphere that holds the seeds.

Take a new look at the flower garden. The lily, iris, and jonquil buds open into hexagonal, six-petaled blossoms. Have you noticed that the green "petals"—the sepals—at the base of a rosebud spread into five-pointed stars as the colored petals unfold spirally? Press a blossom of Queen Anne's lace in a book if you want to see a design for lace as beautiful as any a queen might wear.

Study the fruits and vegetables! Slice a cucumber, and you will find three divisions of seeds. Slice a green pepper, and you will find four divisions. Slice an onion, and the slices will fall apart in circles. Cut an apple crosswise, and you will find the seeds lodged in a five-pointed star.

Best of all, view the seashore with new eyes. If you are collecting shells, examine the spiral on the shell of the tiny sea snail—a spiral that seems to have been drawn there by the huge twirling wave that rolled it up on the beach!

Did you ever notice a sand dollar? It is a very delicate white circle. Look at it carefully, and you will see a little five-petaled flower etched on one side and a larger one extending to the edges of the circle on the other side.

If you have come across starfish in the shallow wash of a

wave, you have surely noticed that most of them are five-pointed. Have you ever seen a sea urchin clinging to the rocks with little feelers like pine needles? Dry it and brush the needles off, and you will find that the small domed shell is marked with a five-divided pattern. When you are about to dive off a dock into the warm August water of the bay, the sight of a sea nettle is not welcome. But pause for a moment and study the lacework and structure of its design before you get your net to scoop it out of the water.

Everywhere in nature are shapes that we call "simple geometric forms." These shapes seem endless. The simple divisions of the circle, by 3, 4, 5, and 6, are infinite in their variety and beauty in nature's constructions. For within nature's law of order there is variation, an over-all order with uniqueness of detail.

Consider the snowflakes. These six-petaled ice flowers are hammered and forged in the upper air by the forces of wind and cold. They are always six-sided—that is their law—yet each one is different. So there is freedom within the law. And it is the study of variations within the law that makes mathematics so fascinating, and by no means easy. For mathematics goes beyond the simple and familiar into the realm of the abstract and the imagination.

These are some of the wonders and laws to observe in nature's showcase. We need to have them called to our attention because we have partly lost touch with nature.

But primitive people were terribly, even frighteningly close to nature and nature's forces. They saw and felt her marvels very strongly. So it was from the vast art gallery of the universe that early man learned the geometry they would use in the dim Stone Age.

3. SECRETS OF THE STONE AGE

Stone Age men—prehistoric men who chipped and flaked their tools from stone—used geometry in two great ways: in *technics*, to make life easier, and in *art* to make it more pleasant. For both uses, they borrowed geometric secrets from nature. They felt a sense of swift motion from the slanting line of a tree when it crashed to the ground. If it fell aslant another tree, they felt its strength as a brace. They felt the force of their own slanting positions as they pushed a heavy rock; they realized that it was easier to push a rock along level ground than it was to push it uphill, and they felt the ease of pushing it downhill.

They noticed the speed of birds in angular flight formation. They noticed the sturdiness of three sticks tied at the top to make a tripod.

They saw that curves had a use, too. A round log rolled.

Early men were beginning to discover *the secrets of line and force*. They used these secrets, which they had found in nature, for countless purposes: to brace their homes, to make their pointed arrows and their wedge-shaped axes, to fashion a log roller, and to extend an inclined plank from the ground to the mouth of a high cave.

Geometry was astonishingly useful—and it was ornamental as well!

They must have noticed that the sky was decorated with stars at night and tumbling clouds by day. The earth was decorated with rolling hills, mountain peaks, winding rivers, flowers, trees, reflections in still lakes, and moving shadows. At the seashore, foam-capped breakers curled and spread fan-shaped along the beach, coming only so far and no farther, accompanied by a rhythmic roar. Birds, flowers, reptiles, fish, butterflies, and insects were patterned with colorful designs. And as they noticed all this, primitive men were discovering *the secrets of beauty* in the symmetry and harmony and variety of nature's forms.

Perhaps unconsciously, they began to borrow these designs as decorations of their own. They borrowed the curve of the snail shell, the repeating pattern of wave after wave, the splash of raindrops and the flash of a flame. They drew them as simple geometric designs, circles and dots, crossed and parallel lines.

And through the millennia, they used these geometric patterns to decorate their persons, their homes, their belongings. They

smeared them with mud and ocher on their own bodies. They carved and etched and wove them in their ornaments and tools, their bowls and baskets. They used them to adorn their huts and temples.

Perhaps you are familiar with such designs on the tepees and blankets and beadwork of American Indians. Did you know that they represent natural phenomena—sun, mountains, trees, lightning? Primitive people everywhere have used these designs and this symbolism.

But their development was a fantastically slow process. In prehistoric times, centuries might pass, after a geometric form was noticed, before it was drawn.

The first geometric form to be admired was probably the circle. Even the earliest men must have observed the circular shape of the huge, red setting sun, the white brilliance of the full moon, the circles in the eyes of their friends, and the spreading circles made by raindrops on a puddle of water or by falling leaves on a still pond.

Admiring a circle was one thing. But drawing a circle accurately was quite another story in those faraway times when men lived *without using circles.*

Today we are constantly surrounded by circles, more than we can possibly count. Look up, look down, look all around you. How many circles can you find in any room? Think of the kitchen, for example. There are circular plates and pans, gas or electric burners on the stove, rims on the glasses. And these are only a start.

During an assembly quiz progam at a certain school, the question was asked: "How many circles are there in the auditorium?"

The first guess was one hundred—far too low. Immediately other students with nimble minds calculated about ten thousand in the circle designs on the ceilings and walls. Then others multiplied the number of students by at least three buttons per jacket, one dime per pocket, four eyelets per shoe, two eyes per face. This came to about twelve thousand. But other contestants knew there were more circles uncounted. They thought of the polka-dot patterns on some of the girls' dresses, the rings in notebooks and the circular holes on loose-leaf pages, and all the nails and screws with circular heads holding the floor and furniture together. Finally they gave up!

So, you see, the circle is everywhere in our lives—sometimes for ornament, more often for use. For nowadays we know the value of an *accurate circle*. Think how important this accuracy is in the tiny mechanism of a wrist watch, or the dials on an airplane panel, or all the "wheels" of modern industry.

Certainly the discovery of how to draw an accurate circle was a great accomplishment for early men back in the Stone Age.

But how was it first done? No one knows for sure. Just who discovered the secret—whether it was a man, woman, or child— was never recorded. We have no record either of when or where the first accurate circle was drawn, with every point on the circle the very same distance from the center. But we can make an imaginative guess as to how it might have happened.

Possibly the first true circle was traced on the ground by an animal tied to a stake; it might run around again and again to the limit of its freedom, and trample a circle on the ground—a true circle of overlapping footprints, all the same distance from the center stake. Perhaps in some such way the "string" first disclosed the secret of the circle.

This circle may have inspired prehistoric children to play a game. It is easy to picture them using a vine for a string, and swinging themselves around, scuffing out with their footprints perfect circles on the ground.

But their parents would probably not be impressed by this early circle game. Men were too busy hunting, and women were too busy tending the fires. This great discovery may have remained unappreciated by the adults for long centuries.

Yet, in the end, the secret of the string was recognized. It too was added to the useful knowledge of line and force, and the circle was added to the beautiful decorative shapes copied from nature. The perfect circle was one more geometric discovery in the slowly growing store of secrets of the Stone Age.

IN THE ANCIENT NEAR EAST

Geometry and Daily Life

4. READING THE SHADOWS

Here our story takes an exciting new turn. The time was just before the beginning of recorded history. The place was the Near East. And geometry was about to change from its primitive beginnings into the ancient practical art of earth measurement.

Some men of the new Stone Age had been through what scholars call a "revolution," because it made such a drastic change in their lives. They were no longer hunters and fishers, but herders and farmers. And in their new roles they learned a fascinating new kind of secret from nature: *how to read shadows, messages from the sun, as a measurement of time.*

As usual, the knowledge came gradually, for it was built upon the distant past.

Men of the old Stone Age had lived through many exhausting millennia, hunting wild beasts for food and being hunted in return. In caves and jungles, across steppes and forests, they had hunted saber-toothed tigers and mammoths and boars—and themselves been stalked and trampled and gored.

Then some wandering groups of men entered the great river valleys of the Nile and of the Tigris and Euphrates. For a long while before this, they had gathered wild berries and fruits and nuts to eat with their meat or fish. But here, by chance, they learned to prepare wild grain with a pleasant taste. And in time, as the climate grew drier, somebody noticed that from grain scattered on the river mud—by the wind, by human hands—a plant would grow. Thirsty animals, crowding toward the rivers, were trapped and penned. Men were starting to *produce their own food* along with great rivers of the "Fertile Crescent."

In these broad, protected valleys, they could water their flocks and cultivate the land. They lived together, season after season, and sometimes they had leisure to plan and dream.

Among those who had most leisure were the shepherds. They watched their flocks under a scorching sun and took shelter in the shade of large rocks. Many times during the day they had to move to keep in the cool protection of the shadows.

Perhaps it was a shepherd, bored by the monotony of his long day, who first placed stones at the end of a shadow as its position and length changed throughout the day.

We can think of him, and other shepherds, as early scientists who discovered how to tell time. If they wandered far from home, they could tell by the shadow when it was time to start back. They discovered also that the shadow pointed out directions on the ground. Along with a need to measure time, they had a need to tell directions. This was before they had any real need to measure distance, for in those dim ages, wandering herders and farmers still moved their homes from place to place, following the wild grasses and the streams.

But *time* was already important to them. As men discussed the usefulness of measuring the time of day, they must have thought about the usefulness of measuring the time of year. At first, the proper seasons to plant, to sow, to reap, to mow, were all guesswork. Many mistakes must have been made, and vital crops lost. What was to be done? If by day the sun marked off the time, why not look to the sky for other needed measurements of time?

Trees bloomed, then birds sang among blossoms and pecked at the first fruit, and what was left developed and ripened. The earth dried up, and the leaves fell. The flood of the river came at repeated intervals. Surely there was a pattern in the seasons. Was there also a clue in the sky to all this?

Remember that these people had little to distract them from the sunlight and shadows by day, from the brilliant stars at night. So they watched for signs.

As the shepherds sat through the nights, the sky produced lively entertainment. A bright parade of star groups marched across the sky, and certain bright wandering "stars" moved among them. In the night sky the moon appeared and grew and disapppeared—and started this interesting cycle all over again. So these early stargazers counted the nights as the moon went through this same routine, and discovered a repeated pattern.

The sun seemed to follow a pattern too. There were intervals when the days grew longer and longer, and the shadows came more slowly. Then this would change, and the days would grow shorter and shorter, and the evening shadows would lengthen early. And this routine, too, would start all over again.

These constant cyclic patterns in the sky and on the earth must have some connection.

In lively conversations around their campfires, the early shepherds must have talked about the constancy of the change in the moon, the shadows, the seasons, and the changing length of the day. Surely the moon and stars, the sun and shadows, were ever-present guide posts for their daily work. The shadows changed through the day, the moon changed through the month, the planets through the year.

So these earliest stargazers counted 30 days while the moon changed; and they watched it change 12 times, from the first bird song in a blooming tree, until the tree bloomed again and the birds returned. They made a crude calendar of 360 days: it can be called the first mathematical formula. The inaccuracies in this first calendar kept men busy studying the sky and shadows for centuries to improve it.

Shadows and how to read them were to be very important in the future science of geometry. But that comes at a later period of our story. For now, you may enjoy reading shadows yourself—understanding their messages about time, just as did the men after the new Stone Age.

It will not be hard, for shadows are pleasant things. They call to mind cool relief from the summer sun. The sun is a master artist. It sketches a moving picture of you as you walk on a sunny day. It sketches the form of a tree and its leafy pattern in dark blues and purples on a bright green lawn.

Furthermore, probably you already know how to read the sun's messages from the shadows. Have you ever had the experience of lying in the cool shade under a beach umbrella and suddenly realizing that the hot sunlight was on your face? You had not moved, but the shade had. Or have you ever wondered if it was time to leave the beach and get home for supper, and then looked at the shadows for a clue? Perhaps you have never thought of making yourself a handy little timepiece. Just prop a stick upright in the sand, and watch its shadow change length and direction, as you go into the water and come back out again, until the shadow finally tells you when to go home.

It is even more fun to study shadows deliberately for their

messages of time measurement. Observe their characteristics and make a real record of them. As you go out on the playground or the ball field, mark the end of a shadow. Just before you leave, notice the change in its length and direction. Find out if all shadows point in the same direction at the same time. If you can remember, mark a shadow at, say 3:30 P.M., and then examine it at the same hour one month later.

Better still, look up pictures of old shadow clocks and sundials. These will show you how shadows were once used.

And shadow reading is still very important today. Trained scientists read aerial photographs. If they are given the date, time, and position of the camera, they can use the shadows to determine, for instance, the heights and intervening distances of the mountains on the moon.

But before the beginning of history, reading the shadows was crucial to human life. For that was the time of the early farmers and herders, when the world's first calendar measured day and night, the phases of the moon, and the returning seasons of the fruitful year.

5. THE ROPE-STRETCHERS

Now at last came the dawn of civilization. And with it, practical geometry got off to a real start in the work of the surveyors or "rope-stretchers."

To build the early civilizations, men had to get along together in settlements. Living in groups meant great advantages and also great responsibilities. The men who lived together in the protected valleys of the Nile and the Tigris and Euphrates recognized this. We have the proof in their ruins.

Each early civilization, as far back as several thousand years B.C., has left on the land the tracings of field markers, irrigation canals, and storage basins. The very earliest tracings go back to about 5000 B.C., the date that archaeologists give to the village of

In the Ancient Near East

Jarmo in Syria, the oldest known village in the world. And many of these ruins reveal the work of skilled surveyors. They are laid out with straight lines and right angles.

Nowadays we take right angles for granted, but have you thought what a big part they play in our lives? If you have flown over farm lands, you probably have noticed that the scene below resembled a patchwork quilt, for the fields are laid out in rectangular patterns. The fences divide one piece of property from another in the same rectangular patterns. And the corners of a rectangle are right angles.

As you drive along through the country, have you noticed that the telephone poles, fence posts, trees, and houses all stand upright and at right angles to the ground? Can you imagine a landscape without the right angle? Try to picture driving at fifty miles an hour, with poles standing every which way. And suppose houses and windows slanted in any direction, and the walls of a room were closed inward. Certainly one's sense of security would be disturbed.

Today we are all so familiar with the right angle that we don't even notice it. Look around the room—how long do you think it would take you to count all the right angles you see? Even in this book, right angles form the four corners of each page. They appear in the structure of the letters L, T, E, and H. In modern times right angles, like circles, are everywhere.

STRING, STRAIGHTEDGE, AND SHADOW

But ancient civilizations did not take the right angle for granted. It was a significant discovery made with the aid of a string. It filled needs that were important to men living together in communities.

One very big problem of the early farmers was to mark off their fields for property rights. They had to protect their own property, and respect the property of others to get along in peace and security and to make progress.

What was the best shape for their fields? Remember, these were among the first fields ever laid out. Men were familiar with the winding shores of lakes, the jagged lines of mountain ranges against the sky, the endlessly varied outlines of the clouds. But irregular boundaries for a field would not be practical.

Ancient men did have a regular pattern to turn to, however. Even in their primitive days, they had known how to weave reeds into mats to cover their earthen floors. The characteristic of the warp and woof in weaving is rectangular. And just as a woven mat covered the earthen floor, so could a rectangular form mark off the boundaries of their fields. Yes, the rectangular form worked in weaving—but how could that form be laid off on the earth?

We know part of the early solution. Pictures have come down to us on tomb and temple walls that show the measuring device

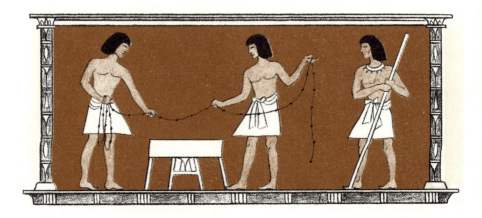

In the Ancient Near East

as a rope. To make a strong rope, men could use a sturdy vine or twist reeds together. And a stretched rope would mark the boundary between two neighbors.

But there were neighbors on four sides, not just on one or two sides. At first, the corners may have been marked off freehand. But men wanted something more dependable than a freehand boundary. The big question was *how to trace an accurate right-angle corner.*

Exactly how this was first done—in Mesopotamia or Egypt—is a question that historians may never be able to answer. But we have some tantalizing clues.

In the wall paintings of thousands of years ago, there are pictures of Egyptian surveyors dragging a knotted rope. From that custom, the ancient surveyors seem to have been called "rope-stretchers." And apparently those ancient rope-stretchers knew that certain dimensions—so many knots or spaces on each side—would make an accurate right triangle.

How did they find out? Perhaps the discovery was made and lost many times. We can only imagine the way it *might* have happened. . . .

Perhaps the day was very hot and some Egyptian rope-stretchers got more tired than usual, laying out a right angle by an old guesswork method. The idea was to make a straight line, and then another crossing it in the middle. This was hard work in the blazing sun—it took three and sometimes four of their measuring ropes with evenly spaced knots.

First they had to lay the straight line. That meant hammering two end stakes firmly into the ground with a knotted rope stretched tightly between. Then they found the middle of the line and drove in a center stake.

Next they took a much longer rope, with plenty of play, and fastened it to the two end stakes. Seizing this rope by the middle, they pulled it as far as possible to one side, opposite the center stake, and drove in a side stake to hold it there.

Finally, they stretched a cross rope from this side stake back over the center stake.

That ought to be a right angle. But the overseer was fussy, so they would have to do the same thing over again on the other side, and then keep at it the rest of the day, until they lined up the ropes perfectly.

In the Ancient Near East

Perhaps at noon when the whole crew stopped to rest in the scanty shade of a clump of palm trees, one man stayed behind, wearily studying the knotted ropes stretched on the ground. Long ago, he might remember, they used to find the middle of a rope by laboriously folding it in two, but nowadays they just used ropes with an odd number of knots and counted spaces to the center knot. Wasn't there some simple way to count for a right angle? He began counting. There it was, *the 3-4-5 right triangle: three* spaces between the knots on one arm, *four* spaces on the other arm, and *five* spaces on the long side opposite the right angle. Naturally, he shouted to the others to come and see. These easy dimensions on a single rope would give them the form they needed so badly, and with hardly any work at all. The rope-knotters could even make a special rope, with large knots already spaced in the dimensions of 3, 4, and 5— and they could peg it anywhere to make a right triangle in a few minutes.

By some such accident, or series of accidents, the string disclosed yet another secret—the perfect right angle.

String, Straightedge, and Shadow

Now that surveyors could mark off accurately, with rope, all the sides and corners of a field, they had ample practice. For every year, when the floods came and receded, the carpet of rich black soil buried the boundary markers, and the fields had to be remeasured.

And the rope-stretchers had another big problem, too. The flood season was followed by a long dry season, for which some preparation had to be made. A network of canals was needed to irrigate the land. Digging them created new difficulties.

If you have ever tried to dig a ditch, you know how hard it is to keep the bottom level, and the sides at right angles to the bottom. But that is important if the ditch is to carry water. Water does not run uphill, and if it runs downhill it flows into one spot. A way had to be found to level the irrigation ditches and straighten their walls.

To do this, the rope-stretchers took two straight sticks of equal length, and spliced them together to form an angle. Then they reinforced this angle with a crossbar, which made a shape something like our letter A. Finally they hung a weighted string from the vertex of the angle.

When the two sticks which formed the angle were on level ground, the string hung straight down through the center of the crossbar. If the ground was not level, the string hung off-center. To make sure that the side of a ditch was vertical, they hung a weighted string against the side. Thus—with level and plumb line—they performed the all-important work of surveying ditches and canals.

And the rope-stretchers helped with one other major problem of community life—taxes.

Because taxes were imposed according to the size of a field, a way of measuring area had to be found. For measuring length, there were two available ways. A rope could be used to measure length: a unit of rope could be doubled by using it twice, or halved by folding it. (A rope with evenly spaced knots was a good measure.) Also, parts of the body could be used as units of length—for instance, the breadth of the finger, the width of the palm, or the span of the outstretched hand from tip of thumb to tip of little finger. And do you remember that the dimensions of Noah's Ark are given in cubits? A cubit is the distance from the elbow to the tip of the longest finger.

But the human body did not furnish convenient units for measuring area. People had to describe the size of a field by a morning's work, or a day's work for a yoke of oxen. But some farmers worked faster than others, and so did some oxen. You can see how disputes would arise. Even at a very early date, man realized that exact measurement was a good way to keep peace.

Again the idea of weaving came to the rescue. In their woven mats ancient men could see a design of little squares. And they could construct a square by making the sides of a rectangle equal. The square became the unit for measuring area.

Naturally, solving all these problems made the rope-stretchers important men in the early communities. For their part, important men were proud to be rope-stretchers.

Thus in ancient Egypt, when the Nile overflow clogged the irrigation ditches with mud and buried the boundary markers, whole villages would go out together to clear them. The leading rope-stretcher of such groups was the local chieftain who supervised the work. To these local chieftains the people paid their taxes with shares of grain or flax according to the size of their property. They also gave them a respect bordering on veneration.

Since the whole civilization depended upon water, which nourished the fields and gave food, these early rope-stretching chiefs were looked upon as divine givers of life. From their number came area chiefs and finally a national leader who was the start of the long line of Egyptian Pharaohs.

For the burial of their Pharaohs, the ancient Egyptians built the great pyramid tombs that are still admired by travelers.

These pyramids are masterpieces of ancient practical geometry. Even today they are enduring monuments to the accuracy of the string-made right triangle and the square, and the early appreciation of sturdiness in the pyramid form.

With the right angle, the pyramid builders laid out accurate direction lines: they accepted the direction in which a shadow always pointed at noon (checked by the position of a fixed star) as the north–south line. By drawing a line at right angles to it, they got an east–west line. These two were always their base lines. The sides of the pyramids' square bases face exactly to north, south, east, and west. And this precision was achieved long, long ago!

The world's oldest man-made stone structure is the step-pyramid tomb at Sakkara. It was built about 2750 B.C. Its square base was pegged off with a knotted rope, In receding steps, it tapered to a peak on top; it was the forerunner of the true pyramid form.

A century later the Great Pyramid was built. A wall drawing has survived that shows us the actual start of the work. It depicts all the pomp and grandeur of the ceremony connected with laying out pyramids and temples in that period. Just as laying the cornerstone of an important government building brings out important officials today, so marking off the ground plan of a pyramid was a great occasion on that day almost 5000 years ago.

The place was selected. The rope-stretchers were on hand. Amidst great crowds, the Pharaoh and his retinue marched to the scene to perform the ceremony. The impressive occasion was called the *Put-ser*, which means "to stretch a cord." The ancient picture shows the Pharaoh holding a golden mallet. And the ancient inscription tells us that the Pharaoh spoke words befitting a royal rope-stretcher:

"I have grasped the wooden peg; I hold the handle of the mallet; I grasp the cord with Seshata [the goddess of the stars]; I cast my face toward the course of the rising constellation; I let my glance enter the constellation of the Great Bear; I establish the four corners of the temple."

6. THE STARGAZERS

Meantime on the flat plain of the Near East, civilization was taking another course. And so was practical geometry, with the accomplishments of the stargazers, who divided the circle and became the world's first systematic astronomers.

We can imagine how the night sky must have looked to the earliest stargazers in the valley of the Tigris and Euphrates Rivers. Those men of old watched the familiar star patterns move across the heavens. They watched the great drama of the procession of the constellations as if it were a gigantic circus parade.

In that starry sky, the ancients picked out figures of human beings and animals and identified them with heroes and gods. They passed on their ideas to others, and we still call the constellations by names that trace back to those times.

And we still use direction-finding tools that derive from instruments they made while stargazing thousands of years ago.

Early peoples of the plain—the Sumerians, the Chaldeans, the Babylonians—needed a guide for their travels and wanderings and wars across that broad flat region. They found it in the stars.

This was a necessity the ancient Egyptians never experienced. Civilizations of the River Nile and the Tigris and Euphrates had progressed in the same direction in measuring off land and building irrigation canals. But different geographical conditions impelled them to develop later on different paths.

The great Nile River was itself a well-defined highway between the settlements of Upper and Lower Egypt. Its narrow valley was protected by mountains and deserts on both sides. For centuries this isolated valley was free from foreign invasion. The people were able to pursue the arts of peace. Wall paintings still exist, after more than 5000 years, that show the Egyptians at work and play—living industrious or leisurely lives in efficient, luxurious, and artistic surroundings.

But the valley of the Tigris and Euphrates, called Mesopotamia, or "The Land Between the Two Rivers," extended over broad, flat plains dotted with little cities. Nomadic tribes roamed over these plains and made war on one another and on the city dwellers. And the many city-kingdoms made war on each

other, too. Wall pictures that have come down to us from this ancient land show warriors, chariots, weapons, and war machines.

In addition to this endless warfare, the peoples of the plain were famous traders. They had no wood or metals, and they needed both for their cities. So caravans of donkeys and camels and flotillas of sailboats set out constantly to exchange goods with neighboring and faraway lands. All this created a problem.

Mesopotamia was so extensive and so bare of landmarks that people had to find a way to lay out directions for their wars and their travels across and beyond the valley. The solution was finally found by their stargazers—ancient astrologers held in great esteem, as were the rope-stretchers along the Nile.

The Egyptians, we know, gave credit for their abundant crops to the rope-stretching surveyors who constructed and maintained the life-giving irrigation ditches. For their part, the people of Mesopotamia believed that if messages from the sky regulated the seasons, they must also regulate the doings of men. They thought that movements of heavenly bodies controlled and forecast important human events, so they gave credit to star gazing priests who studied these movements.

In the Nile Valley people built tombs for their Pharaohs. In the valley of the Tigris and Euphrates, men built temples to their sky gods atop lofty ziggurats (broad tiered towers) for their stargazers. From these high platforms, the priest–astrologers were better able to watch the whole sky, and study and interpret the star movements that directed the affairs of men.

These temples supervised the life of the community. In return, the people presented a portion of their livestock and crops to their stargazing priests, part to be used in sacrifice to their gods and part as taxes to be kept in the temple treasury for the support of the government.

In time, the temples became observatories. And the stargazers became astronomers, and solved the problem of measuring the travels of distant stars.

Already these stargazers had noted the messages of time and direction in the shadows; they had observed the changes in the position of the rising and setting sun through the year. They had

In the Ancient Near East

used the phases of the moon, its periodic growth and waning, to regulate their early calendar. They had watched the steady movement of star-groups across the sky, and the travels of the "wandering" planets in relation to a fixed star.

But they needed an instrument to measure more exactly this brilliant parade across the star-studded heavens. The people needed a measure of direction for travel on the earth. The priests needed a measure for the travels of the stars in the sky. These needs led them to find a very important secret from the circle.

We do not know just when or where an unknown stargazer (or stargazers) made the discovery. The idea of pointing the sides of an angle at two stars must have been very ancient—holding up a string to measure the distance between them was useless, since the length changed as you brought it closer to your eye. But how could you measure an angle?

The answer lay in the division of the circle into six parts. That was the earliest and easiest circular partition, and Egyptians, as well as Mesopotamians, used it for sky measurement. Perhaps we can imagine how it was discovered....

Possibly an old stargazer, looking back on the games of his childhood, recalled tracing a perfect circle on the ground with a string—and used string and stylus to draw a circle on his clay tablet. Now, how would he divide it?

Perhaps he was just toying with his string, or he thought of a field of six-petaled flowers, or he remembered something children did in a game.

After their feet had scuffed out a circle on the ground, sometimes one child stood on the scuff-marks and held the string, while another child went round in a second circle that overlapped the first. The stargazer marked off the curve of his circle

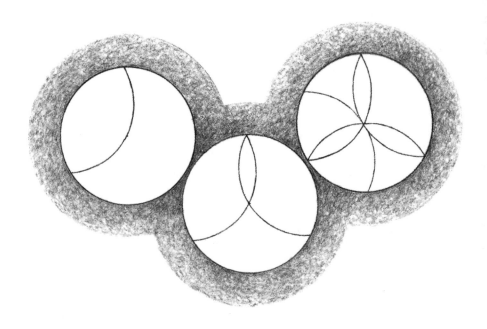

with a series of six arcs, using the same length of string with which he had drawn the circle. When he swung these arcs clear through the circle, he got the flower pattern with its six petals. His discovery was simply that these arcs would cut the circumference into exactly six equal parts.

Anyhow, by chance or by intuition, he hit upon the secret of dividing the circle into six equal parts. After that, it was an easy matter for other stargazers to go on dividing each arc in half, and then in half again. But how many times should they redivide it? The ancient Mesopotamians counted by sixes and tens. Their earliest stargazers thought the year had 360 days. So what could be more reasonable than to continue dividing this circle, with its 6 convenient arcs, into smaller and smaller parts— until there were 360 tiny divisions in all! With these tiny divisions, the stargazers had a new convenient unit of measure:

the arcs on the circumference of the circle would measure the corresponding angles at its center.

As soon as this discovery was made, measuring the star travels was easy. The ancient astronomer attached a movable pointer to the center of a semicircle. With this device, he could follow the planets and measure off their distant travels in units of angular measure on the semicircle.

Using the same device, land directions were easy to indicate, too. He could mark off directions on the earth from the east—that is, from the position of the rising sun on the day when it rose midway between the farthest points where it had risen during the year.

And this unknown stargazer left us a monument to the division of the circle. From that time on, tables of measurements of the star movements were kept in the temples.

As far back as four thousand years ago, these ancient astronomers—with their pointers and semicircles and quarter-circles (quadrants) and sixth-circles (sextants)—observed and recorded eclipses of the moon. But at that remote time such observations were only occasional and unsystematic. Gradually it became the custom to make more frequent observations until in 747 B.C. the series became continuous and a record was carefully kept.

What had started as astrology became a science when men collected the first long series of astronomical observations. It lasted more than 300 years, the longest uninterrupted table ever kept to this day!

Through these records a constant pattern could be seen in the periodic movements of the sun, moon, and planets, a pattern that enabled men to foretell the time of a future eclipse and the future position of these celestial bodies.

Apart from their astronomy, the Mesopotamians left us other great monuments to their division of the circle: the arch and the wheel.

They were probably the earliest people to use the wheel. And when they changed from their older solid wheels to wheels with spokes—the divided circle—they created light chariots for their wars.

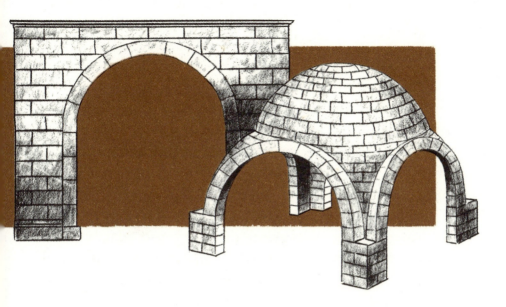

They were also probably the first people to invent the arch.

Nowhere on the broad plains could they find mountains to supply stone, or forests to supply wood. For building material they had to use sun-baked brick. They had to discover a way to use brick as a support for a door or gateway.

Again their knowledge of the circle came to the rescue. They found that if they placed bricks in the form of a half-circle with a wedge-shaped brick in the center, the wedge-shaped brick (keystone) worked as a force against the supporting bricks. Such an arch had the strength to hold the weight of a wall. By interlocking arches, they fashioned a dome. The arch and the dome are still characteristic of the architecture of the Near East.

These circular forms—the arch and the dome—followed the trade routes from Babylon all around the Mediterranean. In time they became the basis of the domes, bridges, and aqueducts of the Roman Empire centuries later. Of course we use them in our constructions today.

So the mementos of the division of the circle, thousands of years ago, still guide our daily life. For many centuries, ships on the high seas—ancient sailing ships, modern steamers—continued to use these dependable signs, the stars. Old, but not old-fashioned, the stars are the same ever-present guideposts that directed nomads across the trackless deserts.

Today, pilots in the air or on the sea are giving up star navigation for the latest radio devices. Yet the very instruments by which they *plot* their courses still reflect the division of the circle by the Mesopotamian stargazers: the compass, divided into 360°—north at 0°, east at 90°, south at 180°, and west at 270°—enables a navigator to mark off in degrees an accurate flight plan.

Likewise, the instruments with which we measure distances on the earth recall the ancient achievement. Global lines of longitude and latitude, spaced through 360°, enable us to place a position on the earth by means of their graph-pattern. And our modern surveyors use a protractor arranged on a transit, so that they can measure distance, direction, and height by a millennial method.

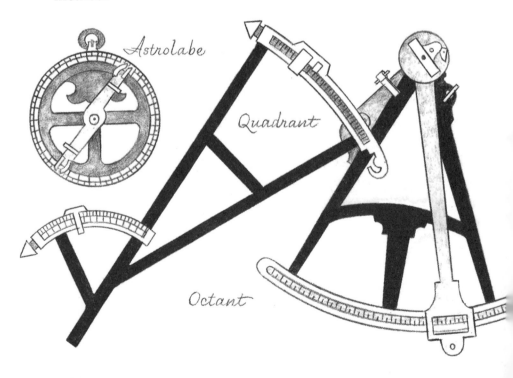

Astrolabe

Quadrant

Octant

In the Ancient Near East

Even our clock is based on the division of the circle. Look at its circular face divided into 12 hours, each hour into 60 minutes, and (if there's a second hand) each minute into 60 seconds.

In these ways and many others, we still use the great accomplishments of Babylonian astronomers after thousands of years. They help take us back in our imagination to that unknown stargazer of long ago—who first divided the circle and made possible the charting of land and sky, the map-making and astronomy, that were the crowning achievements of ancient Near Eastern geometry.

Clock

18th Century Compass Card

Sextant

Small Craft Radar

THE IONIAN GREEKS

Geometry and Thought

7. THE 6TH CENTURY B.C.

At the start of the 6th century B.C., the Mediterranean world was changing. The center of civilization was about to shift to Greece, or Hellas as it was then called. Geometry, too, was about to undergo a radical change—from a purely useful art to a new kind of abstract thinking.

The ancient peoples of the river valleys, the Nile and the Tigris and Euphrates, had done marvelous things with practical geometry. They had used it to lay off their fields and irrigation canals, to construct beautiful buildings and gigantic pyramids, to measure the travels of the stars, to find directions on land and sea.

But by now, both Egypt and Mesopotamia had passed their zenith. Their creative time was over, though they were enjoying a final blaze of splendor.

The Egyptians had developed a sumptuous civilization in their fertile valley, hot and low-lying, with the slow river flowing through it and the limitless desert beyond. Now, under Psamtik II, the land was more prosperous than it had been for almost ten centuries. But Psamtik was a collector of art and antiquities, and his Egypt was hardly more than a museum of past glories.

Mesopotamian culture, too, had flourished between the "two rivers" and across the vast, warm plains. During the reign of Ashurbanipal, most famous of the Assyrian kings, Nineveh had become the largest and most magnificent city in the world. Yet its huge library had also been a collection of learning from the past. And now Nineveh was utterly destroyed, and Nebuchadnezzar was beautifying his great capital city of Babylon. Here he built the famous Hanging Gardens, like a verdant mountain on the flat Babylonian plain, to please his wife, who was homesick for the Median hills. And he encouraged the compilation of star records in the temples. But Babylon's days of power were numbered.

While these ancient centers were basking in revived glory, something fresh and new was stirring farther west.

The Hellenes, a people who streamed down from the chilly forests and mountains to the north, had established their cities on the rocky Greek peninsula. From its cliffs they faced the open sea, and looked out at the snow-capped isles of the dark blue Aegean. It was a craggy land of stubborn soil, where strong men had to work hard to get their food. Soon they were overflowing from Greece proper, taking to their ships and colonizing the islands and the shores of the Mediterranean and the Black Sea.

So the 6th century B.C. was an age of expansion, trade, travel, exploration, and the mixing of the old cultures with the newly awakened one.

During this century of changes the spotlight of ancient history was starting to swing west. For the next 300 years, the mainspring of civilization would be Greece in her creative age. The Greeks were to introduce a new element into culture: reason. Their love of reason would transform art and architecture, philosophy, literature, science, and, in the first place, mathematics.

We have seen what the old Near Eastern civilizations accomplished in *practical geometry* with the help of the circle and the right angle, and how they read the sun's messages in the shadows. Now we shall see that *theoretical geometry* was established by Greeks on these same elements. Its foundations were set firmly, by means of the observation of shadows, on the circle, the right angle, the right triangle, and the relationships within and

The Ionian Greeks

between these forms. The elements were the same, but the approach was entirely different.

This new approach began in the Greek colony of Ionia in Asia Minor, where the brilliant city of Miletus was a crossroad between the East and the West.

The great harbor of Miletus faced west, and welcomed the sailing fleets of Greek and Phoenician merchants. Her rich market was a trading place for the overland caravans of the East, of Persia and Babylon and Egypt. Milesian sailors and merchants traveled to all parts of the known world and came home with strange tales and strange knowledge. And in their teeming city, different races and traditions mixed every day. It was natural that the men of Miletus and the nearby islands should trade not only merchandise but ideas.

These Ionian Greeks were keen and imaginative. They asked questions about everything, and began to collect old answers and frame new ones. Their lively Ionian temperament, their crossroads location, the times they lived in, and the new Greek spirit—all combined to produce a stimulating intellectual environment. Here, in the 6th century B.C., there flourished a remarkable group of individuals. Among them were great poets, and Aesop of "fable" fame. But most fascinating were those whom we would call the world's first scientists.

The Greeks had a different name for them. They were the Ionian "philosophers." Philosophy meant "the love of knowledge," and the term fitted them well.

These early philosophers made discoveries in astronomy, physics, mathematics, geography. Perhaps the earliest was Thales, who studied magnets and measurements. Later came Anaximander, who wrote the first treatise on natural history and made the first map of the world. On the isle of Samos lived Pythagoras, who was credited with inventing multiplication tables (though probably he didn't). But it wasn't only discoveries they cared about. They asked searching questions about the universe. What was the prime substance in everything: was it water or air, was it mind, was it the boundless unknown?

Most important of all, they started a new kind of thinking— *rational* thought: thought based on careful reasoning. The Egyptians and Babylonians had discovered new ways of *doing* things. The Greeks found new ways of *thinking* about them. They observed nature, put their observations in order, tried to find abstract rules.

The first to do this was Thales, whom we have just named as the first Ionian philosopher. He is important in our story for another reason. According to tradition, Thales of Miletus was the founder of geometry.

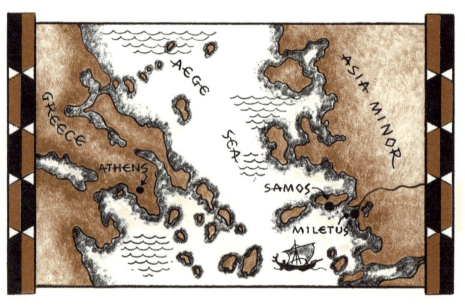

8. THALES AT HOME AND ABROAD

Thales, the "father of geometry," was a sort of Greek Benjamin Franklin. The known facts of his life are few. He was a merchant. He traveled extensively to the older centers of civilization and learned much on his travels. He said "the magnet has a soul because it moves the iron," showing he had studied lodestones. And he is believed to have been the first to experiment with electricity, the static kind in a piece of rubber amber.

But Thales was also a picturesque character and inspired some of the choicest of Aesop's "Fables." Many stories of his accomplishments were told by later writers, some serious and some quite fanciful. True or not, these tales teach us much about Thales' way of thought. He was forever asking "Why?" and working out his answer from what he saw, and standing ready to prove it. Even the anecdotes about his business ventures show this, especially the amusing tale of Thales and the oil presses.

One afternoon as Thales and his friends were discussing money—coins had just recently been invented—Thales made the remark, "Anybody can make money if he puts his mind to it."

His friends immediately said, "Prove it."

Thales was in an awkward spot and he had to think and think. He said to himself, "What item is useful to everybody?" His answer was, "Oil."

In 600 B.C. oil didn't mean petroleum, but olive oil. Olive oil was used for soap. It provided fuel for lamps. It was used for cooking. And it was prized as a skin-softener.

Thales decided to study oil from the tree to the oil press. During this investigation, the first stumbling block he found was the fact that for several seasons the trees had not been producing olives. Why? Thales thought next about weather conditions. In fact he had to do research on the weather of past seasons—the kind favorable to the ripening of the olive, the kind unfavorable.

After that, he also had to try to discover a pattern in the weather conditions, so that he could see what to expect in the future. From his diligent work in laying out the pattern from the past, he calculated that favorable weather conditions were due the next season.

The Ionian Greeks

Now he made a tour among the discouraged olive growers, and bought up all their olive presses. Of course they were delighted to sell them because the presses had been useless for several seasons. Besides, in the past, a grower without a press could always borrow one from a neighbor if the need arose.

But when the big crop came the following year, there were no presses to borrow and none to buy either—Thales had bought them all. So Thales cornered the oil market and made a fortune. Some say he gave the presses back afterwards, because he didn't have time to go into the oil businesss.

Anyhow, the anecdote shows how his mind worked. He was a great observer. He would study the pattern of repeated occurrences and then prophesy the natural path.

Another famous story, told by Aesop, illustrates the same mental traits. It shows that Thales was not above trying to follow the reasoning of a little donkey.

Thales had inherited a salt mine. The salt was transported from the mines by donkeys. They were weighted down with bags of salt at the deposits, and then had to carry them to the market. This donkey train had a long journey in the hot sun.

As they crossed a stream en route, one little donkey was so warm and fatigued that he just collapsed in the cool water and rolled over. Afterward he not only felt refreshed for the rest of the trip, but realized that a great weight had been removed from his back. On every trip thereafter, he repeated this same stunt.

His master Thales was surprised at the beast's fresh appearance, disappointed in his scant cargo, and very puzzled as to how it had been dissolved. For a while the donkey outsmarted Thales, but in the long run Thales paid him back by using some simple deductive reasoning. Thales asked himself, "What sort of thing refreshes the donkey and dissolves the salt? ... A cool stream.... Is there a stream along the route? ... Yes!... What will absorb the water and fatigue the donkey? ... Sponges!" So on the next trip Thales filled the saddle bags with sponges instead of salt, and the little donkey's happy habit was broken.

So as a business man in Ionia, Thales was already using a new type of thinking. But two other interests led him to establish the science of geometry: his travels in the Near East, and his study of shadows.

A story is told about how he took up both.

As he was walking in his garden one night, enraptured by the sparkling splendor of the stars, suddenly the silent stillness of the night was broken by the sound of a great splash and a gurgle. Thales had stepped majestically into a well!

The Ionian Greeks

Fishing him out, his servant remarked with a chuckle, "Master, while you are trying to pry into the mysteries of the sky, you overlook the common objects under your feet."

Nobody likes to be damp and laughed at. In the days that followed, Thales decided to look at the hot dry earth beneath his feet. He would study the shadow patterns that lay there, speaking so eloquently of the sun's messages upon the earth! And he would see more of the earth itself, by traveling to the ancient countries of Mesopotamia and Egypt. (From what we know of Thales, we can guess that he probably decided to engage in some shipping and foreign trade on the side.)

The first stop on his journey was Babylon, a glamorous city with a long history and a large library of cuneiform tablets. There he was fascinated by the impressive records of the stargazers. He stayed for quite a while—poring over the charts, studying the methods of sky measurement, learning the use of the circle and its divisions for measuring angles and directions.

Then he crossed over into Egypt. In that land he mastered the construction of engineering works. He studied the irrigation canals, the well-laid-out fields, the wall decorations showing the history of Egypt in pictures, the designs in Egyptian decorations.

He was absorbing all the old practical geometry of Egypt and Mesopotamia. This was typically Greek. In those times, Greece herself was busily learning from the older civilizations.

And he brought to his travels another trait that was typical of the Greeks and the future civilization they were building. He had a new kind of inquiring mind.

Everywhere he went, Thales studied the shadows traced on these flat ancient lands by ziggurats, obelisks, buildings, and people. He saw these shadows as men had never seen them

before. We might say he had an X-ray eye, because he developed the habit of "seeing through" the obvious to find new meanings—of looking into and beyond visible externals to discover an abstract form and relation.

Here was a remarkable traveler, a Greek Benjamin Franklin indeed! If we can believe the tales of Thales at home and abroad, he took with him his fresh Ionian insight, even as he absorbed the old practical lore of the Babylonians and Egyptians. Out of this combination was to come the new theoretical geometry.

9. HOW HIGH IS THE PYRAMID?

In the Land of the Nile—so the legend goes—Thales amazed and frightened his guides by telling them, as if by magic, the exact height of the Great Pyramid.

The story is worth reviewing in some detail. It shows us Thales' new geometry in action, and enables us to compare it with the old Egyptian kind.

Naturally, Thales' visit to Egypt was not complete without a sightseeing trip to the desert at Giza, to see the three pyramids and the Sphinx half-buried in the sand nearby. In 600 B.C. the pyramids were about 2000 years old.

Thales engaged guides and took a Greek friend along. When they reached those mighty monuments, the guides seemed proud to boast that the Egyptian pyramids had been standing when the ancestors of the Greeks were "long-haired barbarians."

Thales stood for a time admiring the most gigantic of the tombs: The Great Pyramid of Cheops, which covers more than twelve acres! He looked up the great slope, rising to a peak against the cloudless Egyptian sky, and noticed how the brilliant sunlight hit directly against one face and drew a pointed shadow over the desert sands. Then he asked his celebrated question.

"How high is this pyramid?"

The guides were dumbfounded and got into a lengthy discussion. No sightseer had ever asked them that before. Visitors were always content with the dimensions of the pyramid's square base—252 paces along each side. Sometimes the Greek tourists didn't believe that, and had to pace it off for themselves. But this one wanted to know something more: the height. Nobody knew the height of the Great Pyramid. Perhaps, long ago, the builders had known. But by the present dynasty, everyone had forgotten. And, of course, you couldn't measure it. A rope dragged all the way up to the top (and who was going to risk that?) would just give the length of the sloping side. They couldn't think of *any* way to find out the height, short of boring a hole from the top of the pyramid down to its base. But that was impossible.

While the argument went on, Thales and his friend had been walking around quietly, staying close to the pyramid's shadow, where it was cool. Suddenly the Greeks hallooed.

"Never mind my question!" called Thales, as the guides approached. "I know the answer. The Great Pyramid at Giza rises to a height of 160 paces!"

Terrified, the guides flung themselves on their faces before Thales, fully convinced that he was a magician.

To be sure, Thales did not get the answer by magic. He simply measured two shadows on the sand, and then used an *abstract rule* from his new kind of geometry.

To show you just what his method was—and the way he probably worked it out, and how different it was from the old geometry of the pyramid builders—we shall go back and imagine a premliminary scene.

When Thales reached Egypt to spend the winter as tourist and student and merchant traveler, he must have had many things to do. Perhaps there was business to transact on the crowded streets and wharves. Of course he wanted to see the famous monuments, the colossal statues and pyramid tombs.

The Ionian Greeks

But first of all he went to the Temple of Thoth, where the priests were said to have great learning. They proved hospitable, and he spent many days in the cool temple interior, studying the old Egyptian methods.

Now his studies were over, and he was going to start sightseeing.

It was a warm sunny afternoon, and he was sitting outside the temple, waiting to bid farewell to the high priest Thothmes, a most important man. The attendant would take quite a while to fetch him.

As he waited, Thales studied the scene and thought about the accomplishments of Egyptian geometry. In the great empty space before the temple stood a high gilded obelisk, making a fine "shadow pole," or sundial, in the afternoon sunlight. A few white-clad priests and worshipers were standing about, their shadows very distinct, too. Off to one side was the vast temple, perfectly laid out so that its sides would face the four points of the compass. And through its colonnade he could just glimpse a great wall painting, a master piece of Egyptian proportion.

Proportion...It ran, he knew, like a golden thread all through the work of the ancients. Proportion was used by the Chaldean astronomers in the angles of their sextants, and the corresponding arcs of the travels of distant stars. Proportion was used by Egyptian architects, in designing their buildings and erecting the actual structures. But it was best shown in the vast wall paintings with which the Egyptians decorated their temples and tombs. Those vivid scenes were all painted by artists using an innate feeling for proportion.

He had seen them at work. The Egyptian artist had a very

simple means of transferring his small sketch to the huge wall. First he covered his sketch all over with small squares, something like modern graph paper. Then he made squares all over the wall, only large ones this time. Finally he studied where the lines of his sketch crossed the small squares, and then copied these lines in the same relative position over the large squares.

That was intuitive proportion and most practical, the Egyptians at their best....

But as Thales sat there, watching the shadows lengthen outside the temple, he saw something entirely different: abstract proportion.

The Ionian Greeks

Where others saw only the men and the structures, and their shadows in the hot sunlight, Thales saw *abstract right triangles* as well! All these triangles were made the same way: an upright object, a pointed obelisk or white-clad Egyptian; a slanting sun ray that hit the top of the object; and the flat shadow that it cast on the ground.

But Thales saw far more than that. He saw the *motion* of the lengthening shadows. Surely others had seen it too, as they sat waiting, but he saw it with an "X-ray eye."

For as Thales watched, he noticed something truly remarkable. *All the shadows changed together*, in length and direction. At first, they were all half as long as the objects that cast them. Later, they were all the same length as the objects. Later still, the shadows were all twice as long as the height of the objects.

Probably many men had observed something like that, over the centuries. But the Ionian traveler tried to find a constant pattern. He had to prove it was always so, and to find out why.

And he did!

Thales noticed that *all the abstract right triangles changed together, too*—not the whole triangles. The right angle and the height of the object that made its upright or vertical side—these did not change. But the rest of the triangle changed as the sun seemed to change its position in the sky. The sun was so far away that its rays hit the tops of all the objects, and the tips of all the shadows, at the same slant. So, as the sun was higher or lower, the other two angles had to change in all the triangles. And as the angles changed, the other two sides had to change too—*the length of the shadow* (the flat or horizontal base of the triangle), and the *length of the sun's ray* from top to tip (the slanting third side). So at each moment, all the sun-made right

SUN'S RAYS CASTING SHADOWS IN MIDAFTERNOON

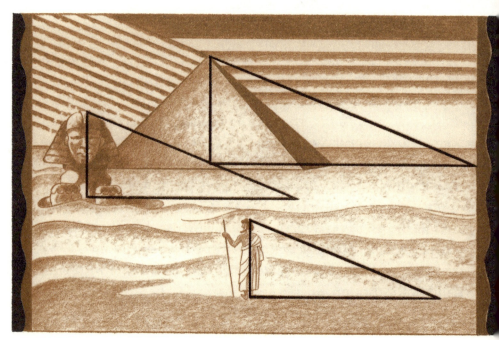

SUN'S RAYS CASTING SHADOWS IN LATE AFTERNOON

The Ionian Greeks

triangles were exactly *the same shape*—not the same size, but the same shape: the right angles and heights of the objects remained unchanged, but the other two sides and other two angles changed as the sun seemed to move across the sky.

Now Thales knew his eyes hadn't deceived him. *The shadow lengths would always change together in the same way, while the heights of the objects must of course stay unchanged.* He had his secret for measuring the height of the pyramid.

Before you hear exactly how Thales accomplished that feat, you may want to try out his secret for yourself.

You can watch these same shadow changes on your own playground or ball field by comparing the right triangles formed by the flagpole, the basketball backboard, and your own height.

Start, say, in mid-afternoon when your shadow is as long as you are tall. At that same time, the flagpole's shadow will be as long as the flagpole is high. And the shadow of the basketball backboard will also be equal to its height. So you can pace off the shadows of the flagpole and the basketball backboard to find their heights, without bothering to climb up with a measuring tape.

Of course, you can start earlier in the afternoon, at a time when your shadow is about *half* of your height. Then you could pace off the other two shadows, double their length, and so find the height of the other objects.

Or you might wait till later in the day, when your shadow is *twice* as long as your height. To be sure, the other shadows would also be twice the height of their objects. So you could pace off the other two shadows, and take half of the distance—and that would give you the height of the flagpole or the backboard.

That's all there is to it, except for one thing more. The shadows won't necessarily be in these obvious lengths—as long, half as long, twice as long. You'll need a simple formula to use with a shadow of any length.

Getting it is easy. The "secret," you already know, is simply a *proportion*. As Thales did, you will always find that at any moment there is a constant relation between one object's height and its shadow, and the next object's height and its shadow. In this case, you are using the equal ratios between the height of *an object* and its shadow, and *your* height and your shadow.

Just put it like this:

HEIGHT OF OBJECT (Ho) IS TO SHADOW OF OBJECT (So) as *Your Height* (Hy) is to *Your Shadow* (Sy)

You can write this as equal ratios, Ho:So::Hy:Sy, or as an equation of fractions, $\frac{Ho}{So} = \frac{Hy}{Sy}$.

Then simply clear the first fraction (multiply both sides of the equation by So), and you get $So \times \frac{Ho}{So} = So \times \frac{Hy}{Sy}$. Since $\frac{So}{So} = 1$, then $Ho = So \times \frac{Hy}{Sy}$.

HEIGHT OF OBJECT = SHADOW OF OBJECT $\times \frac{\text{Your Height}}{\text{Your Shadow}}$

Now that you know the secret for yourself, you will want to see (in your imagination) exactly how Thales measured the height of the pyramid.

When he asked his famous question, the guides, you remember, began to talk and argue. Meantime Thales, who already

The Ionian Greeks

knew the distance along each side of the pyramid's base, 252 paces, was busy pacing off the length of the pyramid's shadow. It measured 114 paces. Thales knew his own height, 2 paces (6 feet). So, just as he finished his pacing, his friend measured his shadow for him: it was 3 paces. Now Thales had all the necessary dimensions; three items of the proportion would give him the missing fourth one, the height of the pyramid.

So he made his calculation as shown in the illustration.

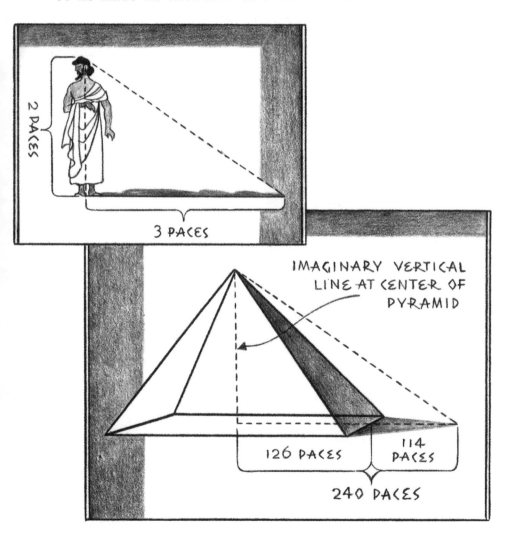

Do you see what Thales did? He used an *abstract right triangle!* He pictured the height of the Great Pyramid as an *imaginary post* from its top straight down to its base. Such an imaginary post would cast an *imaginary shadow*, all the way from where it stood at the center of the pyramid clear out to the tip of the pyramid's real shadow: so the length of this *imaginary shadow* would be one-half the length of the base plus the actual projecting shadow! Therefore:

Height of Pyramid (Imaginary Post) =

$$\text{SHADOW OF IMAGINARY POST} \times \frac{\text{Thales' Height}}{\text{Thales' Shadow}}$$

HEIGHT OF PYRAMID =

$$(\tfrac{1}{2} \text{ its Base} + \text{its Shadow}) \times \frac{\text{Thales' Height}}{\text{Thales' Shadow}}$$

$$\text{Hp} = (126 \text{ paces} + 114 \text{ paces}) \times \frac{2 \text{ paces}}{3 \text{ paces}}$$

$$\text{Hp} = 240 \times 2/3 = 160 \text{ paces!}$$

Of course, the guides promptly spread the news of Thales' magical solution to this seemingly impossible problem. When the priests of Thoth verified that 160 paces was indeed the height of the Great Pyramid, according to the old records, popular astonishment knew no bounds.

The tale traveled far and wide, so far and wide that it has come down to us after 2500 years. And the story has even more meaning today.

For the Great Pyramid was a sturdy monument to ancient practical geometry. But Thales' shadow-reckoning of its height was an even more stalwart monument in the development of reasoning.

10. THE RULES OF THE GAME

With this "new" thinking, Thales was the first to abstract and formulate the *rules of geometry*. By his method, he was also liberating men's minds.

How well his fellow Greeks realized this can be seen in the most famous of all the stories about Thales. It concerns a sensational event that occurred in the midst of a battle.

In 585 B.C., the Medes and Lydians were in the sixth year of a stubborn war. Suddenly broad daylight turned to darkness: the sun gave no light. Terrified out of their wits by the fearful gloom, the warring hosts stopped killing each other and immediately concluded a peace.

History says that this was an eclipse of the sun. Legend says that Thales had forecast it accurately, through the pattern in the records he had studied in Babylon.

Scholars have doubted that Thales predicted the eclipse, but recent studies suggest he used an early imperfect method known to Assyrian court astonomers. Either way, the point of the story is the same. *The Greeks believed Thales had made the prediction, in other words that such a prediction was possible.* Thales' style of reasoning had taught them to look for an orderly pattern in nature, instead of imagining foolishly that an eclipse came about when Apollo, the sun god, hid his face in displeasure.

So great, indeed, was Thales' fame that he was ranked as the first of the fabled Seven Wise Men of Greece. He was supposed to have orginated the motto "Know yourself!" But "Think for yourself!" sounds more like Thales.

He taught men to do just that, with his new abstract rules. Before him, geometry had consisted of isolated observations, ways grasped from trial and error, for handling and calculating material things. Thales showed the need of a careful demonstration, based on a logical sequence of *geometric concepts abstracted from material things.*

To explain how he did this, we shall show Thales actually teaching. There is no evidence that he did. But tradition gives him at least two famous pupils, and he was not a solitary thinker but a busy man of affairs. So he might have expounded his new concepts in some such scenes as these.

When Thales returned from abroad, he probably brought interesting and even expensive curios for his friends, perhaps Mesopotamian seals and amulets, Egyptian jewelry and glass. But the most valuable contribution he brought back to his

The Ionian Greeks

homeland was the wealth of new ideas in his head, the string in his pocket, and the memory of shadows on the sand.

Picture the amusement when his friends would gather around him to hear about his travels, and Thales would take out a piece of string and begin to draw circles on the ground. But not everyone laughed. Some friends were fascinated as he made more circles, and showed how the Babylonians divided them. Or he would knot the string, to show how the Egyptians made a right triangle and squared off their rectangular fields. Or he would reveal his new thoughts about the sun-made right triangles—the abstract ones made by the bright ray, the object, and the shadow.

Soon—we may imagine—a small group assembled from time to time, Thales teaching and expounding, and his friends gradually joining in the discussion.

The met in an open-air place, with white sand on the ground, and poles of different heights to cast different length shadows. Thales would draw circles and triangles and straight lines on the sand, with a string and straightedge, tell in which country he had studied these forms, and explain his new ideas about them. The others would sit and stand and pace in the sun. They would study the diagrams, watch the shadows, swing the string, ask questions, and argue.

It was like an exciting new game: the game of string, straightedge, and shadows.

Every game has its rules. So the eager players would go into huddles to discuss the rules and reasons, and to reach agreements. These rules, Thales showed them, had to be reached through careful step-by-step reasoning, based on agreements and definitions.

String, Straightedge, and Shadow

We can hardly realize how thrilling this new intellectual game must have been to its participants. For the first time in history, men were doing sustained abstract thinking about the principles of line and form. The Babylonians and Egyptians had *used* these right angles, levels, divided circles, geometric designs. Thales and his friends were *thinking* about them—thinking in the abstract, as he had when he measured the pyramid.

The Ionian Greeks

Thales explained the Babylonian division of the circle into 360 degrees and showed how the same measurements applied to angles. Mostly they could see these things for themselves, with just a few definitions and rules:

A circle is a closed curve on which every point is equally distant from a point within called the center. Such a closed curve is called the circumference, and the radius of the circle is the distance from the center to the circumference.

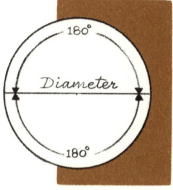

A diameter of the circle is a straight line passing through the center and dividing the circumference into two equal parts, each of 180 degrees.

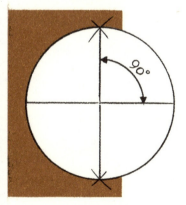

By using a piece of string longer than the radius and swinging equal, intersecting arcs from each end of the diameter, another line called a perpendicular can be drawn throught the center to divide the circle into four equal parts—each of 90 degrees.

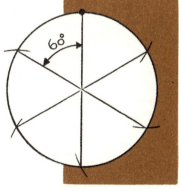

Arcs marked off at successive points on the circumference, with string the same length as the radius, divide the circle into six equal parts—each of 60 degrees.

From these definitions and rules it was obvious that the space surrounding any point may be divided into 360 degrees. Hence a straight line, such as a diameter, drawn through a point is called a straight angle and has 180 degrees.

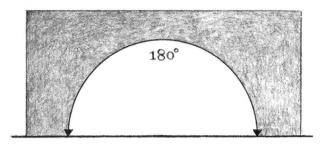

A perpendicular to this straight angle forms two angles—each of 90 degrees, called a right angle.

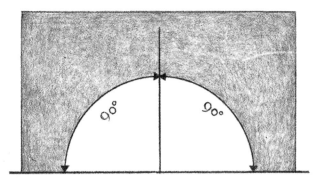

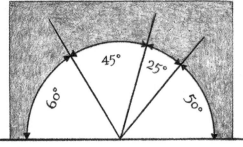

Rule: *Angles adding up to a straight line total 180 degrees, whether they are two right angles, two unequal angles or more than two angles.*

The Ionian Greeks

Speaking of Egyptian rope triangles, Thales reminded his friends that a long rope, knotted at intervals of ten paces, would form a right triangle when stretched on the earth with sides of 3, 4, and 5 units. And so would a shorter rope knotted at one-pace intervals, as would a small cord knotted at intervals the width of a man's hand. The triangles would differ greatly in *size*, but they would be *similar*: each would be a right triangle with sides in the proportion 3:4:5.

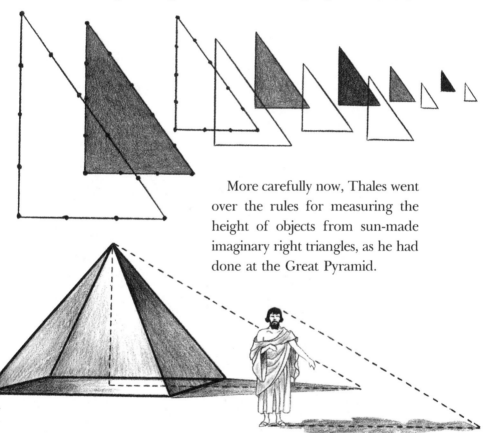

More carefully now, Thales went over the rules for measuring the height of objects from sun-made imaginary right triangles, as he had done at the Great Pyramid.

Rule: *When their corresponding angles are equal, right triangles are similar and their corresponding sides are in proportion.*

85

String, Straightedge, and Shadow

The study of the Egyptian level revealed that the string hanging from the vertex of the triangle is perpendicular to the base and divides the triangle into two right triangles of exactly the same *size*. Then it became clear that any triangle which has two equal sides must have equal base angles. Such a triangle is called *isosceles*, which in Greek means "equal legs."

Rule: *In an isosceles triangle, the two base angles are equal*

The Ionian Greeks

From a familiar pattern, found by dividing the circle into six equal parts, an unexpected new relation emerged. All three sides in each of the triangles are equal to the length of the radius, and all three angles are equal to 60 degrees.

Rule: *In an equilateral (equal-sided) triangle, each interior angle is 60 degrees and the sum of the three interior angles is 180 degrees.*

Since two lines perpendicular to each other form four equal angles of 90 degrees, it was easy to see that, when any two straight lines cross, the two paris of opposite angles are equal.

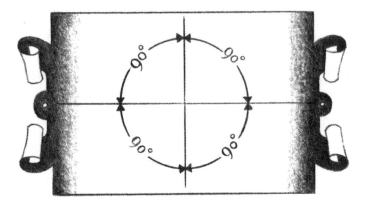

A diagonal line cuts two parallel lines at the same angle, thus forming two pairs of equal angles smaller than 90 degrees, and two pairs of equal angles larger than 90 degrees.

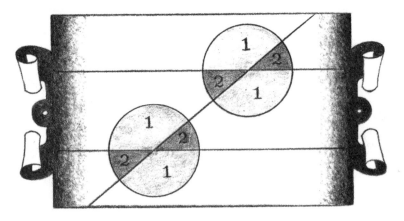

Rule: *The alternate interior angles formed by a diagonal cutting two parallel lines are equal.*

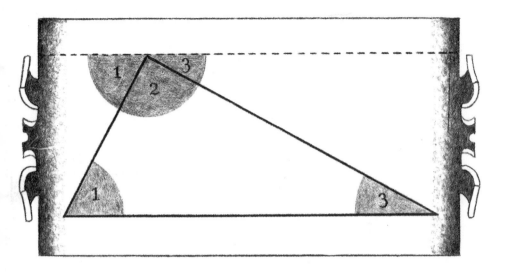

"By constructing a line parallel to the base of any triangle," said Thales, "we can see that the sum of the angles in any triangle is equal to 180 degrees...

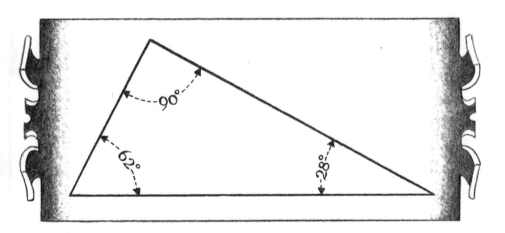

and that in any right triangle, since the right angle is 90 degrees, the sum of the other two angles is 90 degrees."

Finally, Thales showed his friends how to combine these rules to make an important discovery about semicircles. He pointed out that two straight lines drawn from any point on a semicircle to the ends of the diameter form an enclosed angle of 90 degrees. Can you see how he might have done it?

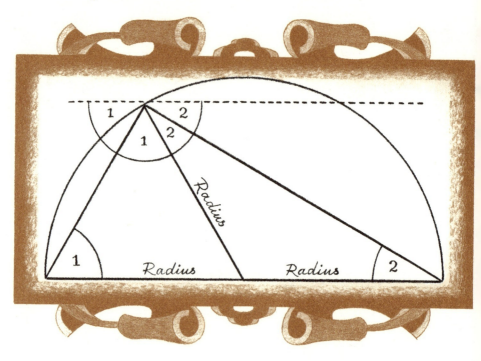

A radius drawn to the point on the semicircle divides the inscribed triangle into two isosceles triangles, since each has two equal sides. We know that each isosceles triangle has equal base angles. And as the inscribed angle is equal to the sum of the base angles, two times this sum must be equal to 180 degrees. Then the inscribed angle is one-half of 180, or 90 degrees.

Rule: *Any angle inscribed in a semicircle is a right angle.*

The Ionian Greeks

In spite of his interest in abstracting rules, Thales was always a practical man, who knew how to apply a rule once he had got hold of it. (Remember the story of the oil presses!) After he had formulated his rules about triangles, he used them to measure the distances of ships out at sea, which was quite valuable for merchants. The achievement was typical of a Milesian.

For he and his Ionian contemporaries were not only philosophers, but inventors or reinventors as well. There was Theodorus, who perfected the Egyptian level into the Greek "diabetes"; and Anaximander, who used the "shadow pole" for astronomical measurements. That was as much a part of Ionian thought as speculation on the basic substance of the universe. And Thales was the first to excel in each branch of Ionian natural philosophy.

So this practical merchant was a true pioneer of rational thinking, in his view of nature and especially his work on geometry. It was he who formulated the very earliest rules for the new game of string, straightedge, and shadows. As the first man in any land to feel the need of such rules, he *made geometry abstract*. And by building rule upon rule, he started its great method of *deductive reasoning*, which would be continued by later geometers.

For the Wise Man of Miletus passed on his rules to others, who were not at all practical—to a strange secret society that would develop them into the foundations of Greek theoretical geometry!

THE SECRET BROTHERHOOD

Geometry, Mathematics and Magic

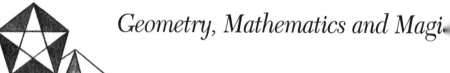

11. PYTHAGORAS AND HIS FOLLOWERS

The early story of Greek geometry is strangely different from its founding in Miletus. Most of what we know is a mixture of myth and magic, shapes and rules, all revolving around the fabulous figure of Pythagoras.

The "divine" Pythagoras—that was what he was called, not only after his death but even in his own lifetime. For the latter part of the 6th century B.C. was still a time of superstition. The Ionian "physiologists" had only tried to find an orderly pattern in nature. Most men continued to believe that gods and spirits moved in the trees and the wind and the lightning. And cults were popular all over the Greek world—"mysteries," they were called—that promised to bring their members close to the gods in secret rites. Some were even headed by seers.

Pythagoras was one of these. A native of the island of Samos, not far from Miletus, he probably had a Phoenician mother and a Greek father, who was a stonecutter. But he gained such a reputation for wisdom and magical arts that people began to whisper that he was son of the god Apollo.

Actually, Pythagoras was Thales' contemporary, for a time at least. He was born about twenty years before Thales died, so his career spanned a later period. Parts of that career are a matter of history. The political situation in Samos became oppressive: a local tyrant, Polycrates, ruled harshly, and the neighboring Persian Empire demanded heavy tribute. So Pythagoras emigrated, as did many other refugees. He settled in Croton, a little island off the tip of Italy. There he founded a famous secret society that contributed a great deal to the development

String, Straightedge, and Shadow

of geometry. We might call it the world's first mathematics club.

But much of Pythagoras' life is enmeshed in legends—not just amusing anecdotes, as with Thales, but wildly fanciful tales. And far too many discoveries are attributed to him. So we must pick and choose our way among facts and fables, in telling the story of Pythagoras and his followers.

To begin with, Pythagoras went on where Thales left off. Let us therefore accept the tradition that he was the older man's student.

Perhaps rumors of Thales' exciting new game of string, straightedge, and shadows had spread throughout Ionia, and people came from the neighboring cities and islands to take part. Anyway, one visitor in particular was attracted to Miletus, to learn this new way of thinking and finding rules and tracing forms upon the ground—the youthful Pythagoras.

The aging Thales must have been pleased with the young man's keen interest; such penetrating questions showed a real thirst for knowledge. Thales taught Pythagoras all that he knew. Then he encouraged him to travel for himself in the ancient lands and study the development of learning at its source.

Pythagoras followed the advice, and his travels were even more extensive than Thales' had been. Fired with enthusiasm by the stories of Babylon, he visited that fabulous city to absorb the learning of the Chaldean stargazers. Naturally, also, he wanted to see the ancient pyramids, obelisks, and temples of Egypt. There he studied the lore of the priests at Memphis and Diospolis.

In addition, he learned a great deal just by traveling to all the known parts of the Mediterranean world. During his long sea voyages, the Phoenician sailors taught him much about the

The Secret Brotherhood

importance of stars in navigation. And like Thales before him, he saw things in a way that men never saw them before.

On the open sea, he realized that the surface of the waters was not flat but curved. He could "see" this whenever another ship appeared in the distance. At first, only the top of its mast was visible over the horizon; then gradually the whole vessel would come into view as it sailed toward them. Surely then, he guessed, *the earth must be round!* And what about the other heavenly bodies?

The moon, when it was full, was a round disk in the sky, rosy or yellowish or silver white. As it waxed and waned, you could imagine that its surface was curved too, and partly in light and partly in shade. Doubtless the moon also was spherical.

String, Straightedge, and Shadow

And the radiant sun itself made a blazing circle in the heavens!

Certainly, concluded Pythagoras, the earth and the sun and the moon and the planets were all spheres. That was the one perfect form: it must be so! In history he is given credit as the first person to spread this idea.

Observing and studying in this way, Pythagoras traveled for many years. Some say he got as far as India and was deeply influenced, for he took up Oriental dress, including a turban. And certain of his mystical ideas, such as number magic and reincarnation, were typical of the East.

Finally he came back to Samos. We don't really know how his countrymen received him, but a number of stories suggest that they were indifferent to all the knowledge he had brought home. This is borne out by the tale of Pythagoras' first pupil.

Tired of finding no one who would listen to his learning, Pythagoras bought himself an audience. He found an urchin, a poor and ragged little fellow, and offered him a bribe. He would pay three oboli for every lesson the boy mastered.

To the urchin this was indeed a bargain. By sitting in the shade for a few hours, and listening attentively to this wise man, he could make better wages than in a whole day's work in the hot sun. Naturally, he concentrated hard while Pythagoras introduced him to mathematical disciplines.

From the simple calculations of the rope-stretchers, to the methods of the Phoenician navigators, to abstract rules and reasoning, Pythagoras led his pupil on. Soon the subjects became so interesting that the boy begged for more and more lessons.

At this point, Pythagoras explained that he, too, was a poor man, and paying someone to listen was getting to be very expensive. So they reached another bargain. The boy had saved enough to pay Pythagoras for his future lessons.

The story doesn't prove that Pythagoras began to collect a following this way. But it shows the fascination of the new game of string, straightedge, and shadows, and forecasts his great role as its teacher.

What we do know for sure is that Pythagoras left Samos and went to settle on a tiny island off the coast of Sicily, which was then swarming with new Greek colonies. This Isle of Croton has an immortal place in the history of mathematics. There Pythagoras finally gathered a group of students around him and founded his famous Secret Brotherhood.

Like other mystery cults of that time, it was a religious order with initiations and rites and purifications.

These "Pythagoreans" had a special way of life. The members —women as well as men—shared all their simple belongings in common. Because Pythagoras taught the doctrine of the transmigration of souls, they were respectful to animals and would eat no meat or fish because in those creatures might live the soul of some departed friend. Nor would they wear garments made of wool, nor kill anything except as a sacrifice to the gods. They bound themselves by great oaths to keep secret all their discoveries and teachings. So devoted were they to their leader that any argument was resolved by using the words of authority referring to Pythagoras: "He himself said it!"

But there was one great trait that set this Brotherhood apart from all the rest. Pythagoras taught that "knowledge is the greatest purification." So his followers were, above all, a study group, bent on gaining the knowledge that would free them from endless rebirths. And to the Pythagoreans—as we shall see —this knowledge meant mathematics!

12. A FAMOUS THEOREM

The most famous thing about Pythagoras is not his Brotherhood at Croton, nor the weird legend of his spending years in a cave and gaining magical powers. It is simply a theorem (or formal rule) of geometry.

The *Pythagorean theorem* says: *In any right triangle, the sum of the squares of the two sides is equal to the square of the hypotenuse.*

This theorem and its proof were a basic advance. It became a cornerstone of ancient geometry and had more influence on theory and more practical applications than any other. Later writers would call it "the measure of gold."

But perhaps Pythagoras ought to be most famous for something else. He was the first to teach mathematics as a liberal education—our very term "mathematics" originated from his course!

The Secret Brotherhood

Pythagoras gave lectures on *mathemata*. In the language of his time, that was the word for studies, but his use of it came to mean mathematics. Pythagorean *mathemata* covered a very large field, but all the parts were interrelated. Perhaps the quickest way to understand this is just to imagine a poster at the entrance to the open-air meeting place where the lectures were held.

This was a curious fourfold range of subjects: music to elevate the soul, numbers and their properties, ancient Babylonian lore about the planets, and the abstract rules of the new theoretical geometry. Each topic was studied from a mathematical standpoint—and the whole course constituted the initiation into the Secret Brotherhood!

The very term "mathematician" meant one who was admitted to the inner mysteries, as distinct from a "hearer" or beginner. "Mathematicians" had to follow a rigorous course for several years, with a stern daily program of meditation, exercise, and study, before they were even permitted to hear Pythagoras intone some teachings behind a curtain. Only after full initiation might they attend his actual lectures.

But the wait may have been worth it. Great mathematical discoveries were attributed to this strange teacher, the most famous being the Pythagorean Theorem. So let us use our imaginations and mingle with the initiates at an exclusive closed lecture. Why not hear the great Pythagoras demonstrate his immortal proposition? Not that he did it in just these words, but these probably *are* the proofs that were used.

Perhaps the session began with an announcement: "I have found at last the solution to a problem that has long been puzzling us." A hush of awe fell on the gathering as "Himself"—in white robe and gold sandals, his head crowned with a golden wreath—took pointer and string and straightedge, and began to lecture.

"Listen to our baffling problem. You older members have already worked on it, but I will review it for the new initiates.

"Here is the *Egyptian right triangle*, the one used by the rope-stretchers, where the sides of the right angle are 3 units and 4 units, and the hypotenuse is 5 units." He drew it on a sandy space, and then added a square on each side, and inner squares. (See illustration on page 101.)

There! You can see, by counting or by calculating the square units, that the total area of the squares on the two sides of the right angle is equal to the area of the square on the hypotenuse"

The Secret Brotherhood

He beckoned to the newcomers, who crowded close, multiplying and counting at the same time:

$$(3 \times 3) + (4 \times 4) = (5 \times 5)$$
$$9 + 16 = 25$$

until all their heads nodded in agreement.

"Now let me show you a Greek design involving right triangles." He drew their attention to the tiled floor on which they were standing, and then traced a similar pattern on the sand, outlining the important parts.

"Here the two sides of the right angle are equal, and the same relation holds." With his pointer, he indicated one triangle and the related squares, and they all counted together. (See illustration on page 101).

"Look! Two triangles plus two triangles equals four triangles. The total area of the squares on the two sides of this right triangle is likewise equal to the area of the square on its hypotenuse."

Again he waited until the newcomers nodded their assent, and then continued:

"In India the priests know other contructions that give similar results; they guard these numbers closely, but we have found some of them. In Babylon, a priestly astrologer whispered to me that there was a secret about this mystery that had never been penetrated."

Now the attention was almost breathless as Pythagoras intoned solemnly: *"That secret is our problem! Would the same relation always be true of ANY right triangle, no matter what the length of its sides, and how could you show this?"*

At this dramatic moment, he withdrew behind a curtain, while attendants played on stringed instruments to indicate an intermission. Pandemonium broke out among the assembled initiates. All the newcomers began talking at once, making suggestions, arguing, and shouting. The older mathematicians, who had worked on the problem themselves, were less noisy but even more excited.

Finally Pythagoras reappeared. Silence instantly fell over the group as he resumed his lecture.

"I will now show you how to construct a wondrous figure which discloses that the answer is always yes! The older 'mathematicians' will realize that by slowly and carefully defining each step of the construction, and using a few simple theorems that you already know, this demonstration can be made into a rigorous proof. Today I will just draw it quickly, so you can all see my great discovery."

He signaled to attendants to smooth the sand, and began to draw, using his pointer to emphasize his words.

"Watch this beautiful construction! I make a square frame,

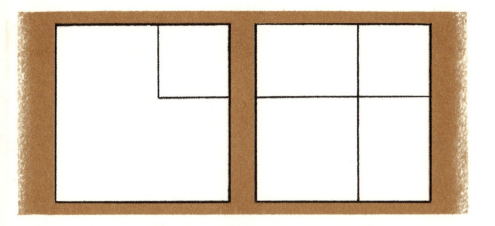

any size, and in its corner I place a small square, any size. Next I draw straight lines, continuing the sides of the small square to the edge of the frame.

"Do you see what my frame now contains? A *small square* and a *medium square,* and *two equal rectangles.*

"And next—we are almost there—I simply add diagonal lines across the rectangles!

"This is the figure I need. My frame now contains a *small square*, a *medium square,* and *four equal right triangles.* Now I will ask you to look more closely at this figure."

Pythagoras beckoned to the attendants, who poured colored sand from jars onto the parts of the drawing, so the pattern showed plainly.

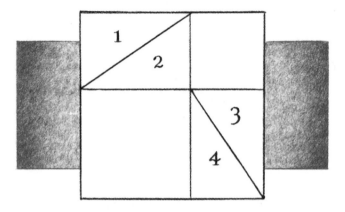

"Look again!" He used his pointer and spoke with care. "All the triangles, you know, are equal; each is the same triangle in a different position. Now, notice how the triangles touch the squares, especially Triangle 2. You can see that the *small square* is the *square on the short side* of the triangle. And the *medium square* is the *square on the long side* of the triangle. So my frame is completely filled by *four equal right triangles* plus the *square on the short side* and the *square on the long side!*"

Pythagoras paused while a low murmur of awe rose from the initiates.

"Now watch!" he intoned. And while they all craned their necks to see, and the attendants poured more colored sand, Pythagoras drew his final masterful figure.

"Watch well! I have only to swing and push these four triangles around, like this, so that they fit perfectly into the four corners

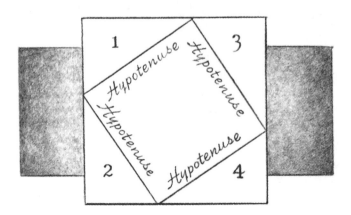

of the frame, and my frame is now completely filled by the same *four equal right triangles* plus *the square on the hypotenuse!*

"Therefore, in any right triangle, the area of the square on one side plus the area of the square on the other side will add up to the area of the square on the hypotenuse!"

A mighty shout—we can imagine—went up from the assembled inner group of the Secret Brotherhood. For this theorem was a true landmark in the development of geometry by the Pythagoreans. Almost all later geometric work involving lengths and measurement was based upon it. And this style of solving problems, especially equations, by *diagraming* them, would remain a chief trait of Greek geometry.

But to the initiate who first heard it, the theorem also partook of a mystical revelation. Tradition says that Pythagoras himself celebrated the occasion by a noble sacrifice—an ox, or a hundred oxen—to his "divine father," Apollo. Some ancient writers dispute this, as the Pythagoreans were vegetarians. Whatever the offering, we can easily picture the festivities described in the verse of legend. Doubtless the "mathematicians" chanted, torches waved, and smoke rose from the sacrificial alter.

> "The day Pythagoras the famous figure found
> For which he brought the gods a sacrifice renowned!"

13. "DICE OF THE GODS"

As time went on, the Pythagoreans made even more exciting discoveries—and gave them strange cosmic meanings.

This curious blend was characteristic of Pythagorean geometry. For the initiates of the Brotherhood were seeking a special key to the universe in this wonderful new realm of numbers and abstract forms: triangles, circles, squares, spheres, and the more elaborate forms they made themselves.

And their search had a thrilling climax. After long and painstaking experiments, they discovered the *five regular solids.* These were remarkable and beautiful polyhedra, or shapes with many faces.

The full tale of these five solids can only be guessed at from

The Secret Brotherhood

bits of legend and history, for all the experiments were top secrets, of course.

To impress this on newcomers, perhaps the first thing they were shown was how to make a mystic "pentagram," the emblem that members of the Order wore on their clothing. By means of a secret device (which we will explain later) a five-sided figure, or pentagon, was traced on cloth. Then its points were connected with diagonals to make a five-pointed star. Finally, around the five points of the star were placed the letters of the Greek word for health, ὑγίεια (hygeia), from which we get the word "hygiene."

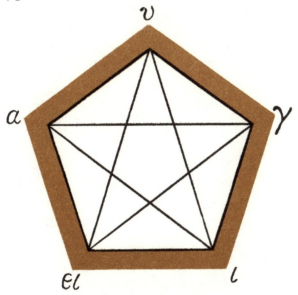

This was the sacred symbol of the Pythagorean Order—the "magic pentacle" that remained a favorite device of sorcerers and conjurors for many centuries. But it was also an experimental discovery: the first known use of letters on a geometric figure.

String, Straightedge, and Shadow

Possibly the next experiment shared with newcomers was a basic one with tiles. Ordinary floor tiles had yielded the easiest example of the Pythagorean theorem. So the Secret Brotherhood went on with a painstaking study of these close-fitting forms that covered many Greek floors.

They made loose tiles of various shapes and placed them in patterns on the ground. And they reached a striking conclusion. There were only *three regular shapes* of tiles that would fit together perfectly to cover a flat area completely: triangles (three sides), squares (four sides), hexagons (six sides).

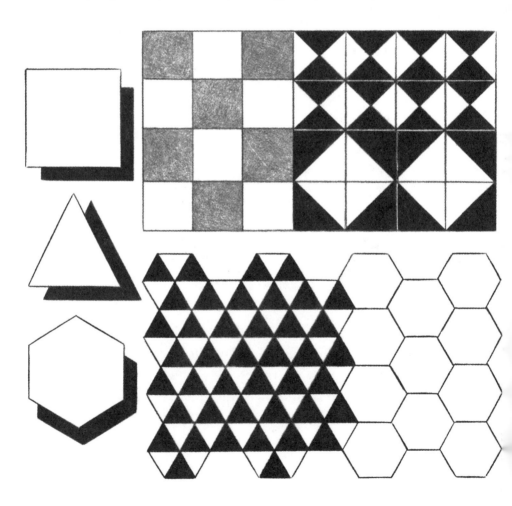

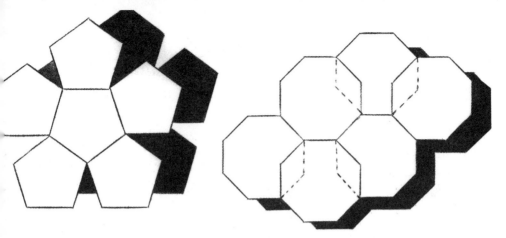

If they tried pentagons, they got a beautiful blossomlike design, but there were gaps between the tiles, and tiles of more than six sides would always overlap. No other regular geometric forms of the same size and shape could be so combined.

They explained this mystery to the newcomers: "Since there are four right angles (360°) around a point, you can only use forms whose corner angles together will make that total. There are just three possibilities: six equilateral triangles with 60° angles, four squares with 90° angles, and three hexagons with 120° angles."

From this simple experiment came the fascinating idea of making "solid angles" by fastening tiles together with mortar, or gluing together shapes of wood, or sewing together pieces of leather. And this led to building shapes with the solid angles.

They called them *regular solids* because all the edges and faces and angles in each solid were equal. And after much experimenting, as we have said, they found five of these solids. The first two had been known from the most ancient times, but the next two were shapes that men had never seen before. As for the fifth, it was such a startling discovery that they thought they had upset the order of the universe!

STRING, STRAIGHTEDGE, AND SHADOW

The Cube. They mortared three square tiles into an angle, and fitted on three more tiles to form a cube with six square faces, which they called a *hexahedron*.

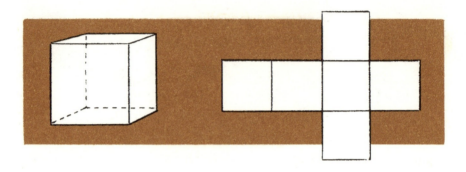

The Regular Pyramid. They put together *three equilateral triangles* into a solid angle, then added one more, to make the base of their four-faced *tetrahedron*.

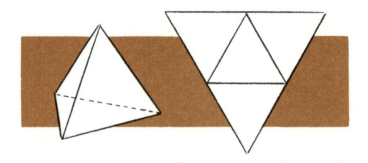

The Octahedron. This was made with two solid angles of four equilateral triangles each, so they gave this eight-faced figure the name *octahedron*

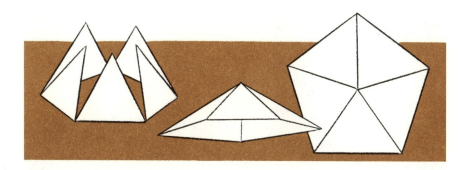

The Icosahedron. Here was a real challenge. When they put together *five equilateral triangles*, they got a surprise. The open base of this solid angle was a *pentagon*. Now they could trace one perfectly for their emblem, instead of just drawing it freehand. (Of course, the device was kept secret.) But how could they make a regular solid, with five equilateral triangles around each vertex? All their early attempts were failures. Finally, someone got the right inspiration—five equilateral triangles for the top, and five triangles for the bottom, and then a center band of ten more triangles based on the old Babylonian pattern. They had made an *icosahedron* with twenty triangular faces.

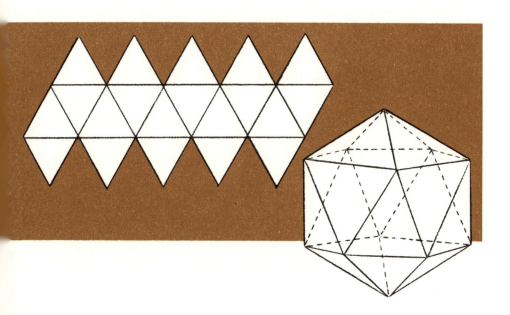

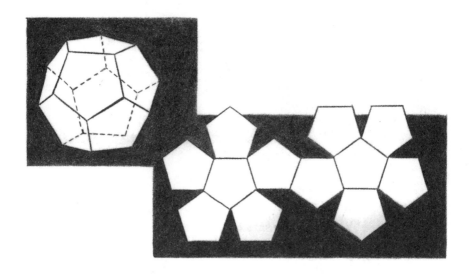

The Dodecahedron. This last form was made with the pentagons so dear to the Pythagorean Order. They used that flower-like pattern of a single pentagon surrounded by five others—the tiles that would *not* fit together on the flat floor. But if the surrounding ones were lifted up, then all six pentagons fitted perfectly in a solid cuplike shape. This could be capped by an inverted one just like it to yield the most difficult and beautiful of the five regular solids: the *dodecahedron* with its twelve pentagonal faces.

These five shapes created a great stir among the first geometers who studied them. Men examined them in fascination and awe, handling them, turning them around in different positions, looking through them as if they were glass. And it was inevitable that the Pythagoreans should finally give them mystical meanings.

The Secret Brotherhood

By that time the Secret Brotherhood had spread to many towns and islands of Sicily and southern Italy. The Sicilian members were friendly with another strange teacher who lived near Mount Etna—Empedocles, who dressed all in purple, gave away his money, and did scientific experiments. Empedocles taught that the world was made of *earth, air, fire,* and *water,* and the first four regular solids came to be identified with these "elements." We know the identification from a famous passage in one of Plato's Dialogues, where it is made by a Pythagorean from Locri in the south of Italy. About the fifth solid, there were many weird stories. Its existence was kept secret as it seemed to require a fifth "element."

The esoteric reasoning, as repeated later, went something like this:

"The cube, standing firmly on its base, corresponds to the stable earth. The octahedron, which rotates freely when held by its two opposite corners, corresponds to the mobile air.

"Since the regular pyramid has the smallest volume for its surface, and the almost spherical icosahedron the largest, and these are the qualities of dryness and wetness, the pyramid stands for fire and the icosahedron for water."

As for the last-found regular solid, with its twelve faces. "Why not let the dodecahedron represent the whole universe, since the Zodiac has twelve signs!"

Such notions were typical of that age. And more than two thousand years later, the famous astronomer Kepler was still so awed by the unique properties of the five regular solids that he tried to apply them as planetary orbits: he assigned the cube to Saturn, the pyramid to Jupiter, the dodecahedron to Mars, the icosahedron to Venus, and the octahedron to Mercury, and he designed a machine to show this! Of course, the attempt was a failure.

Yet even today, these solids seem almost magical in their beauty and their interrelations.

In the first place, it is quite startling that there are only five. An *infinite* number of regular polygons can be inscribed in a circle—their sides becoming so small that they approach the form of the circle itself. But it is not so with regular convex polyhedra inscribed in a sphere. There are only these five possible shapes, and no others.

And these five shapes are connected with one another in a most remarkable way. All five can be fitted together, one inside the next, like the compartments of some magic box. And they are further linked by a strange inner harmony. They can be inscribed in themselves or each other, in certain endless rhythmic alternations. So it's no wonder the five regular solids were long referred to as the "dice of the gods."

14. THE UNSPEAKABLE TRAGEDY

Before the Secret Brotherhood was disbanded, its members really thought they had grasped the key to the cosmos.

Then everything collapsed. Their whole scheme was destroyed by a fatal discovery, and the Order itself was destroyed by traitors and mob violence. Yet as we retell the somber tale, we will find that it was not a complete tragedy after all, for the Pythagoreans did enjoy their cosmic key briefly. This key was not found in abstract shapes alone, nor in music, nor in the stars, but in one factor that—they believed—linked all of these: *number.*

"Himself" had said it: "Everything is number!"

So they followed Pythagoras' teaching that the universe was ruled by whole numbers. That did not mean numbers for ordinary counting or calculating. What interested them was the nature of a number itself, odd, even, divisible, indivisible, and the relations between numbers. This was their *arithmetike.* And they applied it to their other three fields, and found startling number patterns in each.

In music, for instance, a sensational discovery about the relations of whole numbers and musical intervals was attributed to Pythagoras himself.

String, Straightedge, and Shadow

One legend said that on his long voyages he listened to the music of flapping sails, and the wind whistling and whining through the ship's rigging and playing a melody on the ropes. And that he decided then and there to investigate the connection between the tempest of sounds and the vibrating strings.

Another version said that he was strolling through the village of Croton, deep in thought, listening to the musical sounds of hammers striking anvils in a blacksmith's shop, when suddenly, tripping on a taut string that some children had stretched across the street, he got the inspiration for an experiment.

But the most popular story told that the idea came to him straight from the stringed lyre of his "father" Apollo, who was also the god of music.

Anyway, Pythagoras experimented with stretched strings of different lengths placed under the same tension. Soon he found the relation between the length of the vibrating string and the pitch of the note. He discovered that the octave, fifth, and fourth of a note could be produced by one string under tension, simply by "stopping" the string at different places: at one-half its length for the octave, two-thirds its length for the fifth and three-fourths its length for the fourth!

Other musical innovations were credited to him, such as a one-string apparatus for the study of harmonics. But his great discovery was the *tetrachord*, where the most important harmonic intervals were obtained by ratios of the whole numbers: 1, 2, 3, 4. The Secret Brotherhood gave this fourfold chord mystical significance and used to say: "What is the oracle at Delphi? The tetrachord! For it is the scale of the sirens."

And the Pythagoreans even used it for their astronomy. In the relation of number and music, they believed they had found

the pattern that guided the "wandering" planets through the heavens. They pictured the sun and the planets as geometrically perfect spheres, moving through the visibly circular sky on perfect circular orbits, separated by harmonic ratios—musical intervals! Theirs was a vision of time and space revealed in lines, tones, and mathematical ratios. And they even imagined the brilliant planets emitting harmonious tones, the so-called "music of the spheres."

But it was in the connection of number and geometry, their two completely mathematical subjects, that the Pythagoreans were on surest ground.

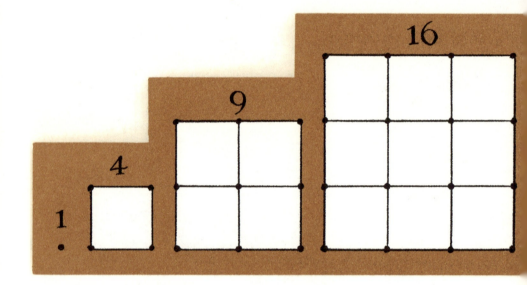

Numbers, they had discovered, *whole numbers*, actually had geometric shapes. There were triangular numbers, square numbers, pentagonal numbers, rectangular numbers, and so on.

This was no wild fantasy, like the singing planets. It was a real mathematical discovery, and came from the circumstance that they did not do their number work by writing the numbers at all. Instead, they placed pebbles on the sand, like the reckoners. But the Pythagoreans placed their pebbles in patterns, adding extra rows for each number. Their two most important series were the *square numbers* and the *triangular numbers*.

The most important number of all, to the Pythagoreans, was the *fourth triangular number,* 10. For it was made up of 1 + 2 + 3 + 4. They called it the "Sacred Tetractys," swore by it in their oaths, and attached marvelous properties to it, as "the source and root of eternal nature."

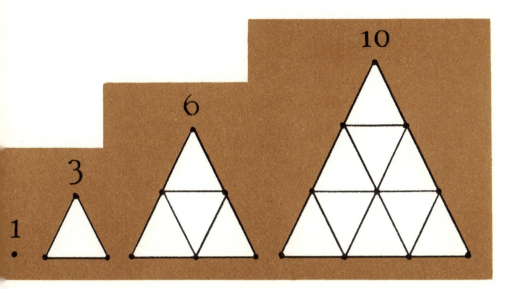

Everything fitted perfectly: the Tetractys, the tetrachord, the four regular solids representing the four "elements," inscribed in a dodecahedron representing the celestial sphere. But it was all too pat, a jumble of luck, imagination, serious mathematical experiments, and old number magic from the East. Just as the Pythagoreans thought they were getting more and more evidence that number was everywhere, the whole system broke down. The entire connection between geometry and number— the foundation of their thinking—was shattered by one disastrous experiment.

Presiding was Hippasus of Metapontum, whose name was to loom dark in the future of the Brotherhood. The idea was simply to find the *numbers* that matched the *sides* of the two right triangles with which Pythagoras had first demonstrated his theorem—the Egyptian triangle and the one from the tiled floor.

Of course, the Egyptian rope-triangle worked perfectly: its 3–4–5 sides made a beautiful Pythagorean series. They indicated the intervals with pebbles. Now what about the right triangle from the Greek tile design, where the two sides were equal?

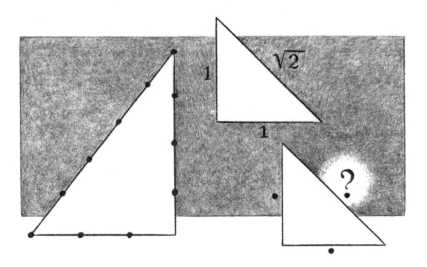

Suppose each side had a length of 1 unit—that would require 1 pebble. Then for the hypotenuse—how many pebbles should they put there? Well, the sum of the squares on the sides would equal the square on the hypotenuse. Therefore,

$$1^2 = 1 \text{ (square on one side)}$$
$$\text{and } 1^2 = 1 \text{ (square on other side)}$$
$$\text{and } 1 + 1 = 2,$$

so 2 is the square on the hypotenuse. And the hypotenuse is the square root of 2.

But what was the square root of 2?

It couldn't be a whole number, since there is no whole number between 1 and 2.

The Secret Brotherhood

Then was it a *ratio* of whole numbers between 1 and 2? Hopefully, they tried every possible ratio, multiplying it by itself, to see if the answer would be 2. There was no such ratio.

After long and fruitless work, the Pythagoreans had to give up. They simply could not find any number for the square root of 2. We write the answer as 1.4141..., a continuing decimal fraction, but they couldn't do that since they had no concept of zero and of decimals. They could draw the hypotenuse easily, but they could *not* express its length as a number. It was "unutterable" —"unspeakable"!

Horrified, the Pythagoreans called $\sqrt{2}$ an *irrational number*. After that, they found other irrationals and swore to keep them secret, for the discovery of these "irrationals" wrecked their entire beautifully constructed system of a universe guided by whole numbers. The breakdown in their mystical morale was followed by the breakup of the Secret Brotherhood itself.

In this final demolition, Hippasus played a decisive role, though his own fate is shrouded in mystery. The Order was already in trouble. Bitter resentment had grown up against its secrecy and exclusiveness, and riots of villagers had driven it out of Croton. Pythagoras himself had died on a neighboring island. And now mobs of "democrats" began to attack the aristocratic Pythagorean societies everywhere.

Against this background, Hippasus took a step that was regarded by the conservative members as sheer betrayal. He broke the oath of secrecy and revealed their most closely guarded discoveries—the dodecahedron and the irrationals. When they promptly expelled him, he set himself up as a public teacher of geometry.

The traitor's punishment was swift and terrible. He was very shortly drowned in a mysterious "accident" at sea, and strange rumors circulated. Some said that a storm had struck his ship as a direct vengeance from the gods: others, that he had been pushed overboard by agents of the Secret Brotherhood. But Hippasus' death was to no avail. The harm was already done to the Order of Pythagoreans, though the discovery of irrational numbers eventually worked for the good of mankind.

The remaining secret groups soon collapsed, torn by outer violence and inner dissensions. And more and more "mathematicians" followed Hippasus' example and came out to earn a living as teachers. Pythagoras' *idea* had been demolished: no longer was there a closed Brotherhood of followers, bound together by a mystical belief in a cosmos ruled by number. Yet his *ideals* lived on in this broader field. He had pursued knowledge for its own sake, loving wisdom for itself. He knew learning could be shared without diminishing, that it lasts through life and immortalizes the learned after death. And the destruction of the Order gave his legacy to the world.

Geometry was now out in the open—and it was the new Pythagorean geometry. True, mathematics was still mixed with some magic: number mysticism, cosmic ideas about the regular solids. But there was, in addition, the famous theorem and its applications, the careful study of shapes, the theory of numbers, and the discover of irrationals.

FROM THE ACADEMY TO THE MUSEUM

Geometry, Art, and Science

15. THE GOLDEN AGE AND THE GOLDEN MEAN

The second half of the 5th century B.C. was the Golden Age of Greece. This was the period of her most beautiful art and architecture, and some of her wisest thinkers besides. Both owed much to the popular new study of geometry.

By the start of the next century, geometry itself was entering its own classic age with a series of great developments, including the Golden Mean. The times were glorious in many ways. The Persian invaders had been driven out of Hellas forever, and Pericles was rebuilding Athens into the most beautiful city in the world. At his invitation, Greek mathematicians from elsewhere flocked into the new capital. From Ionia came Anaxagoras, nicknamed "the mind." From southern Italy and Sicily came learned Pythagoreans and the noted Zeno of Elea. And their influence was felt over all Athens.

High on the hill of the Acropolis rose new marble temples and bronze and painted statues. Crowds thronged the vast new open-air theater nearby, to hear immortal tragedies and comedies by the greatest Greek playwrights. These splendid public works were completed under the direction of the sculptor Phidias and several architects, all of whom knew and used the principles of geometry and optics. "Success in art," they insisted, "is achieved by meticulous accuracy in a multitude of mathematical proportions." And their buildings had a dazzling perfection never seen before—the beauty of calculated geometric harmony.

From the Academy to the Museum

Elsewhere in the city, the impact of the new geometry took another form. On the narrow streets of Athens walked world-famous philosophers, talking to the people, lecturing on mathematics, geography, rhetoric, how to live the good life. Socrates and others asked, "What is beauty? What is virtue?"—and tried to teach men to think out the answers.

Their method was borrowed from the geometers. They called it dialectics, and it was patterned after the deductive reasoning and proofs of geometry. "For geometry," they said, "will lead the soul toward truth and create the spirit of philosophy."

And geometry itself made tremendous strides in the Golden Age and the darker time that followed. Even after Athenian democracy collapsed in the war with Sparta, geometry continued to flourish in the Athens of the restored aristocracy.

But now, in the 4th century, the study was carried on in schools with grounds and buildings of their own. The first and most famous of these was the Academy, headed by the great philosopher Plato. It was located in an olive grove a half-mile outside of town, and over its gate was this inscription:

LET NONE IGNORANT OF GEOMETRY ENTER HERE

Plato's Academy was the earliest institution of higher learning. Its curriculum was frankly inspired by the old program of the Secret Brotherhood. Studies were broader now—the highest branch was moral and political philosophy. But the ideal was still pure wisdom, and the basic training was still in the *"Mathemata."* Plato was partly a Pythagorean.

When his teacher, Socrates, was put to death by the Athenian government, Plato had fled to Sicily. There he studied mathematics under noted Pythagoreans, picked up mystical ideas, and dabbled in aristocratic politics. Finally, he came home to Athens to found his own school and make it the great mathematical center of the Greek world. Most of the mathematicians of that era were his friends or associated with his Academy.

Perhaps the most gifted geometer to study there was Eudoxus of Cnidus, who finally broke the deadlock of the irrationals, and freed geometry for the great advances that were to come. How he did this—with his work on the Golden Mean and his new theory of proportion—is an exciting story. And if we add a bit of imagination, it gives us a fascinating glimpse of Athens and the Academy in Plato's time.

At the age of twenty-four, Eudoxus came to Athens from his home town of Cnidus on the Black Sea, in order to study at Plato's Academy. He was so poor that he could not afford lodgings in the city, but lived in the small seaport of Piraeus and walked to school every day. Of course, he had already studied some geometry; it was the entrance requirement. But at the Academy he got particularly interested in the matter of an irrational number on a geometric figure. For in Athens the problem was in plain sight every day, in a concrete, or rather, a marble form.

On the high Acropolis, against the shimmering sky, stood the beautiful temple called the Parthenon—the most wonderful monument of the Age of Pericles, the "perfect" building whose ruins enthrall us even today. The Parthenon had been designed by Ictinus and Callicrates according to mathematical principles. Its surrounding pillars were an example of "number" applied: 8 pillars in front, an even number, as Pythagoras had advised, so no central posts would block the view: but 17 pillars on each side, where it was all right to have an odd number. And some of

its lines were deliberately curved and slanted to correct optical distortions.

But above all, the Parthenon was a crowning example of proportion in architecture. Scholars still marvel at the logical and harmonious ratios in the whole building and its various parts. And this beauty was achieved with one of the "dynamic rectangles" then in vogue.

Like many Greek temples of the time, the Parthenon used the "root five rectangle," a rectangle with an irrational side, the square root of 5. How did this "root five rectangle" come to be used? How was it constructed and shown to be irrational? How did Eudoxus analyze in it the most beautiful of all linear proportions, the Golden Section, or Golden Mean? That is our story.

The development was natural in the architecture of the Golden Age. Greek builders, we must remember, did not have a minutely graduated measuring rod, in inches or centimenters, like ours. Ground plans were still laid out in the old way, with string (rope), straightedge, level, and carpenter's right angle or "set square." And some of the older temples, and even a few new ones, were quite carelessly designed.

But as geometry became popular in Athens, architects took to drawing careful plans with string and straightedge, for geometric constructions could be enlarged easily and accurately in the building itself.

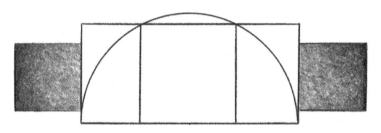

Temples remained severely rectangular, but now the favorite rectangle was made by a "construction": a square inscribed in a semicircle. This figure gave you the shape of the rectangle: it was as *long* as the semicircle's diameter, as *high* as the inscribed square. Calculating its numerical dimensions was easy with the Pythagorean theorem; any builder or Academy student knew how in those days. The rectangle had an irrational dimension. *When its width was 1, its length was $\sqrt{5}$.*

This "root five rectangle" was enough to discourage any member of the Secret Brotherhood—but Eudoxus belonged to a new age. After studying for a while at the Academy, he went to Egypt, where he studied under the learned priests. Afterward, he traveled and established his own school. Then, years later, he returned to Athens to revisit his former master Plato.

From the Academy to the Museum

This time, Eudoxus arrived not as a poor student, but an acknowledged master of geometry. In token of his importance, he now wore his beard and eyebrows shaved in the Egyptian manner. He was accompanied by several of his own disciples. A holiday in his honor was declared at the Academy. All the students wanted to see him, and they crowded into the famous open-air lecture space shaded by the grove of olive trees. And there—we may imagine—he gave them the geometric solution of the proportion in the "root five rectangle," which had puzzled him as a student.

Eudoxus chose his words with care. He had promised to tell Plato a great discovery at dinner, and it would be based on this novel demonstration.

"I will ask you," he said, "to disregard numbers entirely, and forget all about the numerical dimensions of the 'root five rectangle.' We will try instead to find a proportion among the *geometric* quantities. So now, look at the construction itself, the square inscribed in the semicircle." Using string and straightedge, Eudoxus drew it on the sand.

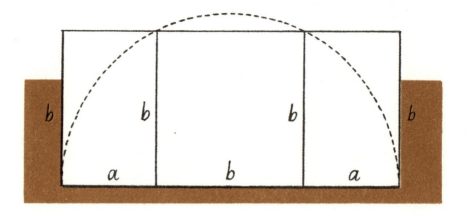

"Look at the straight lines in the whole construction. You will see that there are only two geometric quantities throughout. What are these? One is b, the width of the square, wherever it occurs. Now study the *diameter* of the semicircle. On that line there are three segments. The *long segment* is simply b, the base of the square. The two *short* segments, a and a, on either side, are equal—for each equals the radius minus $\frac{1}{2}$ the base of the square.

"Reduced to its simplest terms, therefore, the problem is to find a *proportion between the geometric quantities* a *and* b, irrespective of any numerical dimensions. So here is the figure once more, simplified to show the problem in this simplest form.

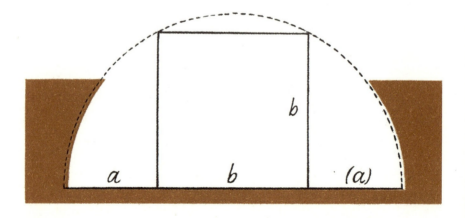

From the Academy to the Museum

"Consider only the line

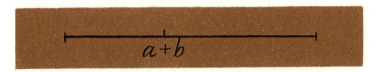

that is, only that portion of the diameter, where our two quantities a and b can be defined as *a short and a long segment of one line*.

"Now here is my question: *What is the proportion that links* a *and* b, *the short and long segments of this line?* Does anybody see how to find out?"

A ripple of excitement rose from the students gathered in the grove of the Academy, as they peered at the diagram and discussed this "simple" problem in whispers. Plato himself stood by, smiling. Finally, when no one volunteered, Eudoxus raised his hand for attention and continued.

"Nothing could be simpler than the answer. It involves a very easy construction that you all know already. From the upper right corner of the square, I will just draw two lines to the ends of the diameter. What does that give you?" He pointed to an eager student in the front row.

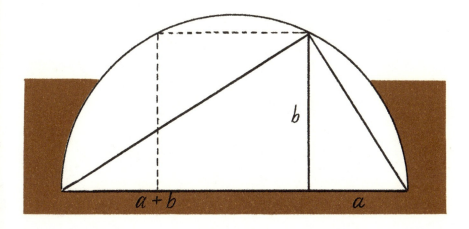

"A *right triangle,* of course," the boy almost shouted. "Lines drawn from any point on the circumference to the ends of the diameter make a right angle!"

"What else do you see?"

Several students answered at once: "Inside this large right triangle are two other right triangles! They are formed by a side of the square—but it is now a *perpendicular line* dropped from the *vertex* of the large right triangle to its *hypotenuse.*"

"Absolutely right!" said Eudoxus, pointing them out. "We will call one S for Small, and the other M for Medium; and the large right triangle, of course, can be L for Large. Now, do you see any relationship between these three right triangles?"

There was a pause, while all the students stared intently at the diagram. Suddenly a boy called out from the back row, "The three right triangles are similar, aren't they?"

"How do you prove that?" Eudoxus was nodding his approval.

"Sir, they are similar because their angles are equal. If you will kindly spin the three right triangles around and draw them side by side and upright, then everyone else can see the proof."

Eudoxus gladly obliged, and, using his pointer, he explained for the benefit of the slower students, "Notice on the figure that each of the smaller triangles has an angle in common with the large triangle. But we know that in any right triangle the sum

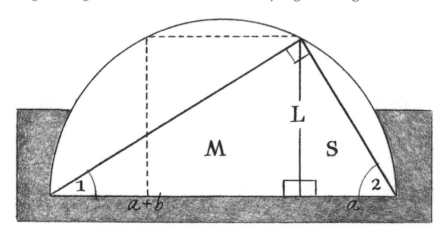

From the Academy to the Museum

of the other two angles is 90°. So each of the remaining angles must be equal respectively. The three right triangles are therefore similar, just as—what is your name, lad?—just as Meno here has said, because their angles are equal. Meno has solved the problem!"

"But sir," protested Meno in amazement, "of what use is it for us to know that the right triangles are similar?'

"Of what use?" repeated Eudoxus, laughing. "Look again, all of you, and you will see the beautiful proportion that links the geometric quantities a and b." He pointed in quick succession to all the drawings on the sand.

"Just take the dimensions from the final figure, and put them on the easy-to-see similar right triangles, just the Small and Medium ones. You know that when right triangles are similar, their corresponding sides are in proportion. Therefore,

Short Side of Small △ is to *Long Side of Small* △ as

Short Side of Medium △ is to *Long Side of Medium* △

or in other works: a is to b as b is to $a + b$

"That is your proportion! Just read it off on the line, and you will see how beautiful it is:

$$a:b = b:(a + b)$$

THE SHORT SEGMENT IS TO THE LONG SEGMENT AS THE LONG SEGMENT IS TO THE WHOLE LINE!"

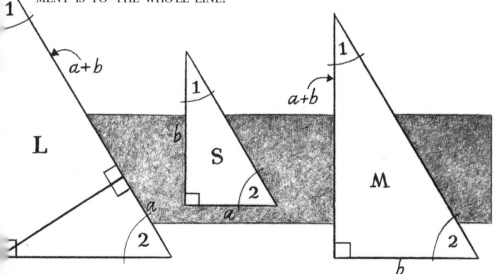

"High-rete! High-rete! High-rete!" cried all the students in unison—the Greek equivalent of three cheers.

Plato himself joined the chorus of praise and made a short speech: "You have just seen a beautiful demonstration and proof —one of the most ingenious in geometry. This proportion is far more significant than the problem that led to it. So I will ask you all to review the construction for your next assignment.

"By inscribing a square in a semicircle, you can do something truly marvelous with a straight line. You can divide that line into two unequal sections, in such a way that the short section is to the long section as the long section is to the whole line. Do you appreciate this proportion? You are thus dividing a line geometrically into its extreme and mean proportional. This section, or cutting, of a line is so important that from now on we will call it *THE SECTION*." Plato drew and wrote on the sand.

Plato gave a banquet in Eudoxus' honor that night—history records the event—and heard the rest of the discovery from Eudoxus' own lips. Before we join them at dinner, let us pause (like the boys at the Academy) to appreciate the importance of "The Section." Plato himself, in his writings, always called it that. But later writers named it the Golden Section or the Golden Mean.

The lasting fame of the Golden Mean rests not only on the sheer beauty of the proportion itself, but on its use in architecture and art. The "root five" and Golden Section rectangles were used frequently in Greek buildings. Scholars have since found that many of the loveliest classical vases and statues cherished today, on the hills of Greece and in museums throughout the world, are based on this same section. And sculptors and painters down the ages have continued to make use of it.

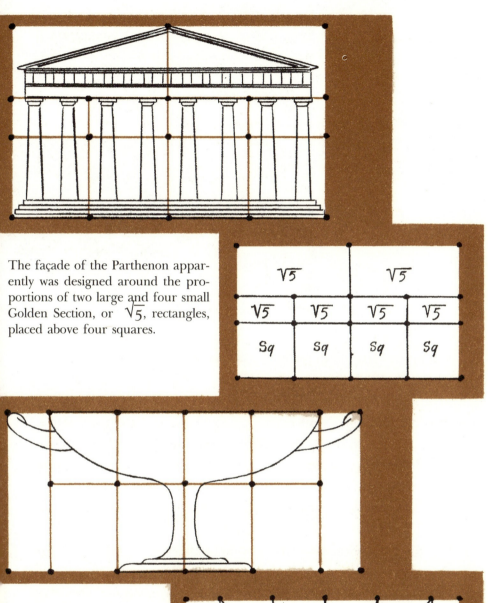

The façade of the Parthenon apparently was designed around the proportions of two large and four small Golden Section, or $\sqrt{5}$, rectangles, placed above four squares.

This Greek vase, known as a *kylix*, was designed to be contained within a Golden Section. The bowl of the vase follows the proportions of four squares placed together horizontally.

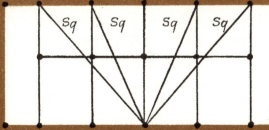

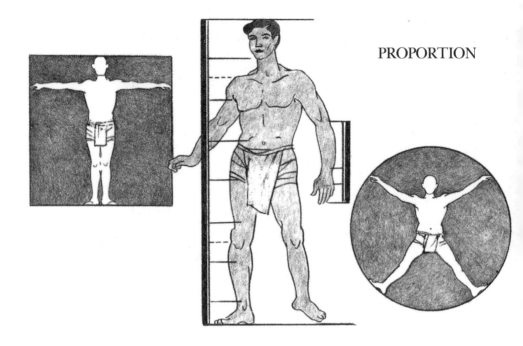

PROPORTION

The Golden Mean is a surprising clue to the proportions of the human body. Just look at the different lengths in your own hand and fingers and forearm, and you can see this yourself. The length of the first finger joint is to the length of the next two joints as those two are to the length of the whole finger! The length of the middle finger is to the length of the palm as the length of the palm is to the length of the whole hand! The length of the hand is to the length of the forearm as the length of the forearm is to the whole length from fingertip to elbow.

Experts have made many more measurements, and have found that this proportion runs through the whole human skeleton—not exactly, of course, but as a kind of "ideal" proportion or standard of beauty. That is why the Golden Mean has fascinated some of the greatest artists through the centuries. Leonardo da Vinci, for instance, called it the "Divine Section."

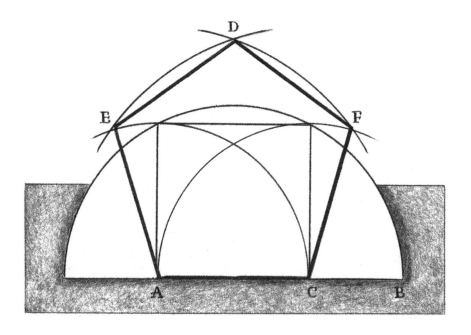

Taking the line *AB* as radius and using *A* and *C* as centers, draw arcs intersecting at *D*. Using *AC* as radius, draw arcs that cut the long arcs at *E* and *F*. Then *AC, CF, FD, DE, EA* form a pentagon, A five-pointed star can be formed by drawing AF, EC, DA, and DC.

Yet the immediate use of the Golden Mean in geometry, during Plato's time, was even more remarkable.

The Section was actually the key to the geometric construction of the pentagon and of the fifth regular solid, the dodecahedron, with its twelve pentagonal faces—not their mere freehand drawing or building up with tiles as before, but their perfect construction with string and straightedge. These and other beautiful shapes can be drawn easily if you just use the Golden Section.

"But the most important thing about The Section," said Eudoxus to Plato at dinner, "is *the kind of thinking* it stimulates. In The Section, the length is irrational, yet it presents no difficulty because it is handled *geometrically*. So I have been working on a new definition of proportion—extending the idea of number to include the irrationals, and the idea of length so theorems will be correct for all lines...."

Of course, this conversation is imaginary, but Eudoxus of Cnidus actually was the greatest mathematician of his age, and the Golden Section theorems were his most striking achievement. Another great geometer, an Athenian friend of his and Plato's, named Theaetetus, probably worked on the theorems first, and Plato himself doubtless taught the subject at the Academy. But it was Eudoxus who finally broke the tyranny of number, with his magnificent new theory of proportion, so we have made him the hero of our story. He really must have contemplated the Parthenon in his student days. And we used "poetic license" to let him demonstrate The Section in the olive grove on the real occasion of his return visit to Athens. That way, you could see for yourself The Section's brilliant role in the Golden Age of Greece —and how, at Plato's Academy, Eudoxus and others freed geometry for the still more brilliant developments to come.

16. A ROYAL ROAD, AFTER ALL

During the 4th century B.C., Greek geometry burst its bonds and went on to the tremendous discoveries of the "age of giants." And Greek culture, too, burst from the mainland of Hellas and spread to most of the eastern Mediterranean.

Both developments were connected with the romantic figure of Alexander the Great. After Plato's time, teachers and alumni from the Academy had gone on to found schools of their own. In particular, Plato's most famous associate, the great philosopher Aristotle, had set up the Lyceum in Athens, and started the systematic classification of human knowledge. And Aristotle's most renowned pupil was the warrior king Alexander of Macedon, who tried to conquer the world.

In thirteen years, Alexander extended his rule over Greece proper, and Ionia, Phoenicia, Egypt, and the vast Persian domains as far as India. Then he died, and his empire broke up. But throughout those far-flung lands, he had founded Greek cities and planted the seeds of Greek civilization—the Greek language, Greek art, and, of course, Greek mathematics.

Mathematicians traveled with his armies. And there is even a story of how hard Alexander himself worked at the study of geometry.

Menaechmus, who had studied with Plato and Eudoxus, was trying to teach Alexander some geometric proofs. The lesson went badly. "Master," exclaimed Alexander, "in my kingdom there are royal roads built smoothly, as short cuts for the king. Can you not make this task easier for me?"

Menaechmus made the now famous reply, "Sire, there is no royal road to geometry."

Yet there was to be a royal road, after all. And before and after Menaechmus, many geometers labored mightily to build it.

During the 5th and 4th centuries B.C., these geometers held the Greek belief that abstract thought, not mechanical work, was the proper occupation for a gentleman. So they kept aloof from practical applications, and devoted themselves to smoothing and polishing existing proofs and finding more and more new ones. These thinkers were of various kinds—dreamers, adventurers, discoverers, organizers. Some fitted into one or another of these categories, some belonged to all of them. We shall only call the roll of the most important. Among the earlier ones were Archytas of Tarentum, Plato's geometry teacher, Hippocrates of Chios, who tried to fit together all the rules, and Theodorus of Cyrene, who discovered many of the irrationals. In Plato's own time, the two greatest were Theaetetus of Athens and Eudoxus of Cnidus. And at the Lyceum, and at the Academy after Plato, many more worked to improve the definitions and assumptions and make new abstract discoveries—including Menaechmus, Alexander's taskmaster.

But after Alexander's conquests, the antimechanical taboo was lifted. A new atmosphere prevailed in the Hellenistic or part-

Greek, monarchies of his successors. World trade, improved navigation, better farming methods were the order of the day. And mathematicians turned their attention to practical matters. They made water clocks, irrigation devices, cogs for launching ships, various sorts of tackles and gears.

The most illustrious mathematician of this later age was also the mechanical wizard of antiquity, Archimedes of Syracuse. Among his many inventions was the so-called "screw of Archimedes," a device for pumping water from the holds of leaky ships or draining flooded mines.

So vast was his knowledge of leverage that he once said, "Give me a place to stand on, and I will move the earth!"

But Archimedes' most fabulous accomplishments were in connection with the defense of Syracuse, where he was attached to the court of King Hiero II. That rich and beautiful city was a tempting prize to Marcellus, who plied the Mediterranean as head of the Roman fleet. Time after time, he directed attacks on Syracuse, but each assault was repulsed by the ingenious machines of Archimedes.

On one occasion the Roman ships were burned by fireballs hurled from a catapult behind the city walls.

Another time, a large claw maneuvered by levers and pulleys actually grabbed the prows of the Roman ships, raised them in the air, and flung them back into the sea.

At another attack, a huge mirror concentrated the sun's rays and focused them on the Roman fleet, setting ships on fire.

Naturally, at each suggestion of a return to Syracuse, the unhappy Roman sailors would cry, "Oh no! Not again!" And Marcellus himself used to call Archimedes "that geometrical hundred-armed giant."

Archimedes' many machines proved the value of science in war and peace even at that early date. Yet he did much more important and far-reaching work in pure mathematics. He and the other Alexandrian geometers were every bit as devoted to abstract thought as their predecessors. In fact, from the Golden Age through the Age of Plato and right into the Hellenistic Age, they were all absorbed in a set of fascinating abstract questions.

These were the *three famous puzzles*—constructions that had to be performed using only straightedge and string. One was called "squaring the circle": how could you construct a square with the same area as a circle? Another was "trisecting an angle": how could you divide an angle into three equal parts? Another was "doubling the cube": how could you construct a cube whose volume would be double that of another cube?

About the origin of the cube puzzle, a curious story was told. It seems that a great plague ravaged Athens in 430 B.C., and the

citizens appealed to the oracle at Delos for help. The oracle replied that the plague would be stayed if the Athenians would double in size the altar of Apollo without changing its shape. The alter was a cube!

Historians do not think this tale is true. Rather, they believe, it was made up later to hide the fact that the "three geometric puzzles" were really useless problems. But working on what may seem useless has frequently been the task of mathematicians, and such tasks, pursued with care, patience, and persistence, have led to most useful results. A whole book could be written about useful results from useless problems.

In the case of the three geometric puzzles, they were not only useless but quite impossible with only those tools. These constructions simply cannot be made with string and straightedge alone! But more than two thousand years elapsed before that was definitely proved.

So from the 5th to the 3rd centuries B.C., many geometers worked in vain on the three puzzles. And attempts to solve them led to the invention of new curves that broke the rules—they were made by a variety of mechanical or three-dimensional means.

Menaechmus himself, struggling with the problem of doubling the cube, dreamed up the idea of cutting a cone with a plane, The sections he cut were an epochal discovery, and later on Archimedes worked on them himself.

They were in the shape of three new curves, the *ellipse*, the *parabola*, and the *hyperbola*—the Greek names of that age are still in use today, and we still call them *conic sections*, because they can be cut from a cone.

If you cut a cone of your own, you can see these curves for yourself. Take an ordinary ice-cream cone—not a real one, as it would crumble if you cut it, but the shape of one. Set the vertex up; this is a *right circular cone*. If you cut straight across it, parallel to the base, you will get a circle which is also a conic section. Slice clear across it at an angle, and you get an ellipse. Now, just slice off one side of the cone, cutting parallel to the slanting outside of the cone itself (the line of generation), and you get a parabola. Finally, slice straight down parallel to the vertical axis of the cone—and stand another cone upside down on top of the first one and slice it in the same way—and you get the two curves of a hyperbola. Picture the sides of your cone extending endlessly, and you will see the open arms of the parabola and the hyperbola reaching out toward infinity. These three curves—the ellipse, the parabola, and the hyperbola—are a perfect example of remarkable results from a mathematical puzzle.

In the next century, the conic sections were studied by

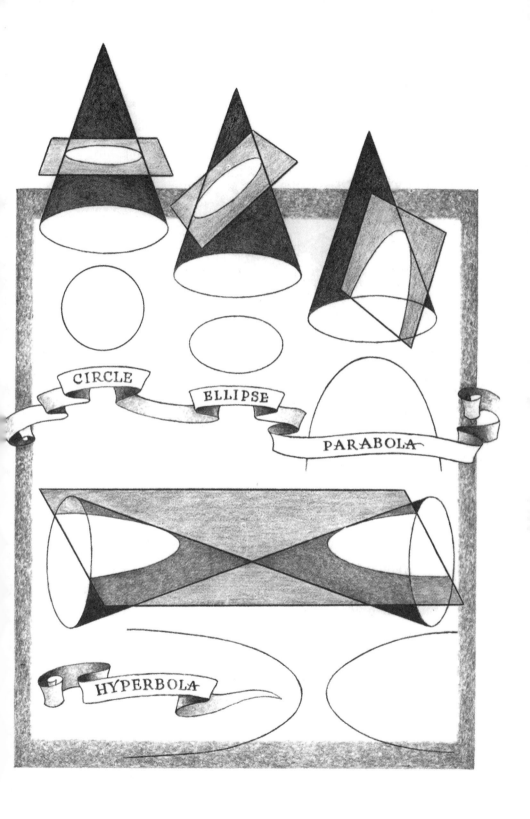

Archimedes, the famous Apollonius of Perga, and others, mainly out of sheer interest in geometry.

But it took two thousand years for the conics to be recognized for what they were: the paths of motion for all bodies, celestial and earthly alike.

It wasn't until the 17th century that the great astronomer Kepler discovered his famous law that the path of each planet around the sun is actually an *ellipse*. In the same century, Galileo proved that a cannon ball or any other missile shot into the air will travel a path that is a *parabola*. Less than a century later, Newton developed his universal laws of motion and his great law of gravitation. Without the *Conica* of Apollonius, which he knew thoroughly, it is unthinkable that Newton could

have formulated these basic laws of modern astronomy and physics.

And today, twenty-three centuries after Archimedes studied them, you can see the conics everywhere, for they have countless uses in science and industry.

Every time you throw a ball, the ball's path traces a parabola. When a jet of water rises from a fountain, it describes a parabola as it falls back into the pool beneath. The path of Astronaut Shepard's trip into outer space and back was a parabola. And this curve has characteristics that make it valuable for reflecting light and sound waves. That is why parabolic reflectors are used in searchlights, flashlights, car headlights, and in radar antennas and radio telescopes.

As for the hyperbola, it is the path that the sun's shadow of an object traces on the ground in the course of a year, and its applications are equally wonderful. This curve is used in LORAN (Long Range Navigation), a radar-like system that enables a pilot to set and hold his course in any weather. He does this by means of radio signals and a map with Loran "lines of position," hyperbolic curves.

So the conic sections—found by Menaechmus from a "useless" puzzle—have proved tremendously useful to later science and industry. And so have the great practical achievements of the Alexandrian geometers.

Archimedes' work on levers and floating bodies was the beginning of the science of mechanics. And his method of determining curved areas and volumes was the forerunner of Newton's calculus. In fact, both the theoretical and the applied mathematics of the Hellenistic Age prepared the way for the massive achievements of Sir Isaac Newton, who laid the foundations of the modern exact sciences. As Newton himself said, "I could not have seen so far, had I not stood on the shoulders of giants."

So Menaechmus was partly wrong, after all, that there was no royal road to geometry. For the giants of Greek geometry did build a royal road that others would tread after two thousand years.

And this "age of giants" was to produce a pinnacle of its own.

17. THE WHOLE, ROUND EARTH

Our story of ancient geometry reaches its thrilling climax at the start of the 3rd century B.C., for then a famous geometer wrote down the whole *theoretical* subject in the best-selling mathematics text of all time, and soon after, another geometer performed the most spectacular *practical* feat. He used a shadow to measure not just a pyramid, but the whole round earth!

These two events took place in the new Greek capital of the age-old land of Egypt.

Founded by Alexander the Great and named for him, Alexandria had become the leading metropolis of the ancient world. By now, it was a flourishing royal city, a beehive of commerce, the most important seaport on the whole Mediterranean. And Alexandria was also the world's main clearinghouse for ideas.

This sumptuous cosmopolitan city was the gathering place for the best scholars and scientists of the age. Savants from many lands made their discoveries in the "museum"—a graduate school that carried on studies in literature, medicine, astronomy, and mathematics. The accumulated learning of the past was stored in the great annexed Library, with nearly a million books on scrolls.

The librarian was a Greek named Eratosthenes. A universal mind, he was a mathematician, a specialist in history, an astronomer, and a poet besides. And around 250 B.C. he did something almost incredible in those times. Eratosthenes measured accurately the girth of the planet he lived on!

Strange as it seems, he had a practical purpose in mind. As a great geographer, he understood that the earth was round, and he was mapping the known parts of it. On his map of the world, Eratosthenes put all the data and distances he could get. The project was typical of that era, when the Mediterranean was becoming "one world" for the first time. It was one world of scientists; astronomers, mathematicians, and geographers in many lands were pooling their knowledge. And it was one world of trade, of ships and sailors, who needed maps. To make his map more accurate and useful, Eratosthenes wanted to determine the width of a degree of latitude. But for that, he had to know the circumference of the earth. How was he to measure it?

The inspiration came to him one day as he was traveling up the Nile on a summer study trip. He noticed with excitement that on the longest day of the year the noonday sun shone straight down a well at Syene, a town about 5000 stadia up the river from Alexandria. He could see the shape of the sun reflected on the surface of the water at the bottom of the well. But from there northward to Alexandria where he lived, the sun never got directly overhead.

So Syene was really on the Tropic of Cancer! To a geographer that was most important, and he explored the region in order to draw the Tropic on his map.

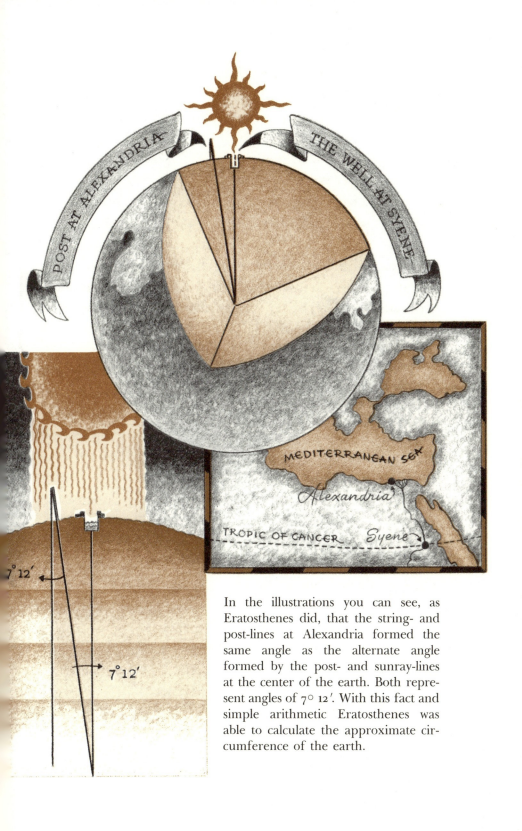

In the illustrations you can see, as Eratosthenes did, that the string- and post-lines at Alexandria formed the same angle as the alternate angle formed by the post- and sunray-lines at the center of the earth. Both represent angles of 7° 12′. With this fact and simple arithmetic Eratosthenes was able to calculate the approximate circumference of the earth.

But the sight of the sun in the well fired his imagination even more. Just that single observation, plus his knowledge of geometry and his own active brain, told him how to determine the distance around the earth. He did it by means of a shadow and some remarkably shrewd deductions. Eratosthenes simply took the known distance between Syene and Alexandria, due north—as reported by camel caravans and professional "step-counters"—and then measured a single angle at the right place and the right time.

He made his historic measurement at Alexandria, at noontime on the longest day of the year. At that moment, he knew, the sun was shining straight down the well at Syene, 5000 stadia away, casting no shadow. But at Alexandria, where he stood, an upright post was casting a shadow.

So Eratosthenes stretched a string from the top of the post to the tip of its shadow. Then he measured the angle between the post (at its top) and the string. Do you see what he had done? The string represented the sun's ray that was casting the shadow. *He had measured the angle at which the post met the sun's ray: It was a 7° 12' angle.*

Now look at the picture on page 151, where we have cut away a piece of the earth, and you can see the brilliant deduction that Eratosthenes made from this angle.

He just assumed an imaginary line from the sun rays at Syene continuing straight down through the well to the center of the earth, and an imaginary line from the post at Alexandria continuing down to the center of the earth, too. Of course, he assumed the sun's rays were parallel. So at the center of the earth, the post line met the parallel ray line at the same angle that the post met the ray at Alexandria! (When a transversal cuts

From the Academy to the Museum

two parallel lines, the alternate interior angles are equal.) *So Eratosthenes made the alternate angle at the center of the earth $7° \, 12'$. The rest was easy.*

This angle goes into $360°$ just 50 times. (There are 60 minutes to the degree.) Then, since the length of the arc made by $7° \, 12'$ was about 5000 stadia, the distance around the whole earth must be 50 times that, or about 250,000 stadia.

In Egypt there were about 10 stadia to a mile, so his measurement was around 25,000 miles—very close to the actual circumference of the earth, as it was measured in later centuries.

Eratosthenes' estimate was the most accurate in ancient times, and the climactic feat of the ancient practical art of geometry, or earth measurement. Yet his contemporaries thought of him as a "second-string man," for around the same time, and in the same city of Alexandria, there lived another geometer whose name is more widely known than any mathematician's in history.

Surely you have guessed who it was. None other than Euclid himself! His masterpiece, the *Elements,* became world famous in his own lifetime, and nowadays it is still just as famous. For more than two thousand years, ever since he wrote it, students have been studying elementary geometry from his great work. The chances are that it is the basis of your geometry book today.

The incidents of Euclid's life are unknown. But we can infer the traits of painstaking accuracy, imagination, dogged determination, and above all logical thinking, that led him to assemble and organize everything that had been accomplished in geometry up to his time.

Of course, there had been previous attempts by other writers. But it was he who finally arranged the whole subject in the complete and orderly outline that was desperately needed.

Earlier Greeks had struggled to find, in simple geometric forms, some basic rules of line and construction and relationship for a clear and accurate description of nature. Afterward, proofs and ideas grew like mushrooms, with more and more theorems established and problems solved. By this time, there was a pressing need for one summary that could be accepted and used by Greek geometers and their students from Asia Minor to Sicily.

Euclid of Alexandria wrote that texbook, but it was far more than a mere textbook. The *Elements* was also a work of art. Out of the pieces of a mathematical jigsaw puzzle, he created a clear and beautiful picture. He traced geometric facts, logically and clearly, from the very first principles that Thales had found. He built these facts, one upon the next, into a truly magnificent edifice. We have spoken often of mathematics in the arts. But the *Elements* shows us the reverse, art in a work of mathematics.

It was as though Thales and the Pythagoreans had quarried great marble slabs from nature, and through the centuries that followed many minds had carved and polished each piece until at last the whole was put together by Euclid into a simple and perfect structure as lovely as any Greek temple.

In addition, the *Elements* told the history of an age. For in making his compilation, Euclid had carefully collected, put together, and reported the work of the preceding centuries. His book contained most of the important discoveries of the Greek masters of the classical period. Here was an indispensable collection of beautiful and useful definitions and theorems. But the arrangement surpassed even the content.

From start to finish, Euclid's *Elements* was held together by rigorous deductive reasoning. From a few wisely chosen axioms (agreed-upon common notions), he had gone on, step by step,

From the Academy to the Museum

to the most advanced proofs. Perhaps no other human creation has shown how so much knowledge can be derived from reasoning alone. So through the ages, his book has been used as a training in logical thought, not just for mathematicians, but for philosophers, theologians, logicians, lawyers, and statesmen. Seekers of truth in all fields have studied and imitated its form and procedure.

But these are just high lights of Euclid's unique accomplishment. By writing the *Elements,* by putting everything together is lasting logical form, he transmitted intact to posterity almost the whole of Greek geometry.

With Euclid's immortal *Elements* we close our tale of geometry in the ancient world. That story has brought us all the way from primitive hunters to scholars at the Alexandria University. We have seen ancient men, using the early practical art, gradually learn to tell time and direction, to lay out their fields and dig irrigation ditches, to design and decorate their dwellings and temples and tombs, to record the travels of the sun, moon, and planets. We have watched the merchant Thales working out the first abstract rules, and the mystical Secret Brotherhood developing these rules, and studying shapes, and trying to link geometery with numbers. Finally, we have seen the later Greek geometers influencing philosophy and art of the Golden Age, and laying the foundations of Hellenistic and modern science. From start to finish, it was a great adventure of the human mind. And all of it, from Stone Age men to Eratosthenes and Euclid was achieved with just the string, the straightedge, and the shadow.

INDEX

Academy, Plato's, 125, 134, 138, 140
Acropolis, 124, 126
Aegean Sea, 60
Aesop, 61, 64, 65
Alexander the Great, 139, 140, 149
Alexandria, 149, 150, 151, 155
Alternate interior angles *see* Angles, alternate interior
Anaximander, 91
Angles, 9, 50-56, 83-90; alternate interior, 88; base, 86, 90; central, 53, 84; measurement of, 53, 83; right, 8, 17, 37-41, 45, 73, 84, 89, 90, 94, 153; solid, 109; straight, 84; *see also* Triangles
Apollo, 80, 105
Apollonius of Perga, 146
Arch, 54, 55
Archimedes of Syracuse, 141-48
Archytas of Tarentum, 140
Arcs, 52, 53, 83; *see also* Circle, divisions of
Area, measurement of, 44, 105
Aristotle, 139
Ashurbanipal, 59
Asia Minor, 61
Assyria, 59
Astrologers, 50, 54, 102
Astronomy, 50, 53, 54, 57, 62, 71, 94
Athens, 124-38
Axioms, 154

Babylon, 59, 60, 61, 62, 67, 94
Babylonian astronomy, 57, 62
Babylonians, 48, 57, 60, 62, 68, 81
Black Sea, 60

Calendar, 34, 35, 51
Chaldean astronomy, 71-94
Chaldeans, 48
Circle, 8, 17, 20, 22; divisions of, 51-57, 84; in design, 24-27; properties of, 83 *see also* Semicircle
Circumference, 53, 83
Clocks, 35, 57, 141

Comets, 17
Compass, 13
Conic sections, 144-48
Constellations, 46, 48
Croton, 93, 97, 98, 116, 121
Crystals, 17
Cube, 110, 113, 142, 144

Definitions and rules, 83-87
Delos, oracle at, 143
Dialectics, 125
Diameter, 83, 84, 90, 128-32
Direction, 8, 9, 31, 45, 49, 50, 56
Divine Section, 136
Dodecahedron, 112, 113, 119, 121
Dome, 55
Donkey, story of Thales', 65-66
Doubling the cube, 142, 144

Earth, 19, 22, 95; circumference of, 151-53; measurement of, 9, 44, 51, 149
Egypt, 39, 44, 59, 61, 67, 70, 94
Egyptians, 39, 44, 45, 48, 59, 62, 67, 68, 82, 90, 91
Elements, 113
Ellipse, 9, 146
Empedocles, 113
Equilateral triangle, *see* Triangles, equilateral
Eratosthenes, 150-55
Euclid, 10, 153-55
Eudoxus of Cnidus, 126-34
Euphrates River, 29, 36, 47, 48, 50, 59

Fertile Crescent, 29
Fire, 16
Flowers, 18, 19, 22, 52
Franklin, Benjamin, 63, 68

Galaxies, 17
Galilei, Galileo, 146
Giza, 70
Golden Age, 124-38
Golden Gate bridge, 8

Index

Golden Mean, 124-38
Golden Section, 134, 135-38
Gravitation, 146
Great Pyramid of Cheops, 45, 69, 70, 78, 85
Greeks, 9, 59-61, 62, 67

Hanging Gardens of Babylon, 59
Harmony, 12, 14, 16, 22, 114, 124
Hellas, 59, 124, 139
Hexagon, 17, 108, 109
Hiero II, 141
Hippasus of Metapontum, 119, 121, 122
Hippocrates of Cheos, 140
Hyperbola, 9, 146
Hypotenuse, 98, 100, 101, 120, 121, 132

Icosahedron, 111, 113
India, 139
Indians, American, 24
Inscribed triangle, *see* Triangles, inscribed
Ionia, 61, 66, 94, 139
Irrational numbers, 121, 122, 126, 138, 140
Irrigation, 36, 42, 44, 48, 50, 59, 67, 141
Isosceles triangle, *see* Triangles, isosceles

Jarmo, Syria, 37

Kepler, Johannes, 113, 146
Keystone, 55

Leonardo da Vinci, 136
Level, 43, 82, 86, 91, 128
Lightning, 15, 16, 24, 93
Lines, 8, 16, 21, 22; *see also* Parallel lines; Perpendicular lines; Angle, straight
LORAN, (Long Range Navigation), 148
Lyceum, 139, 140
Macedon, 139
Mackinac straits, 8
Marcellus, 142

Mathemata, 99, 125
Mediterranean, 59, 60, 94, 150
Menaechmus, 140
Mesopotamia, 39, 48, 50, 52, 59, 67
Meteors, 17
Miletus, 61, 91, 93
Moon, 8, 24, 32, 34, 35, 51, 95; eclipse of, 54
Mount Everest, 8
Music of the spheres, 117
Musical intervals, 115

Navigation, 13, 56, 95, 141, 148
Near East, 29, 47
Nebuchadnezzar, 59
Newton, Isaac, 146, 148
Nile River, 29, 36, 48, 59, 69
Nineveh, 59
Noah's Ark, 44
Numbers, 118, 119, 155; see also Irrational numbers

Obelisks, 67, 71, 73, 94
Octahedron, 110, 113
Oil presses, 64-65, 91

Palomar, Mount, 8
Parabola, 9, 144, 146, 147
Parallel lines, 88, 89, 152
Parthenon, 126, 135, 138
Pentacle, magic, 107
Pentagon, 17, 18, 107, 109, 111, 112
Pentagram, 107
Pericles, 126
Perpendicular lines, 83, 84, 86, 88, 92, 132
Persia, 61, 139
Pharaohs, 45, 46, 50
Phidias, 124
Philosophy, 61, 125
Phoenicians, 61, 94
Planets, 17, 51, 96, 117
Plato, 113, 134-37
Plumb line, 43
Polycrates, 93

Index

Polygons, 114
Polyhedra, 106, 114; *see also* Solids, regular
Prism, 17
Proportion 14, 71, 72, 78, 85, 124, 130, 135; *see also* Golden Mean; Golden Section
Protractor, 56
Psamtik II, 59
Put-ser, 46
Puzzles, 9, 142; *see also* Squaring the circle; Doubling the cube; Trisecting an angle
Pyramid 17, 45, 69, 70, 76-78, 85, 94, 110, 113; *see also* Great Pyramid of Cheops
Pythagoras, 62, 93-121, 128
Pythagorean Brotherhood, 95, 121
Pythagorean theorem, 98, 105

Quadrants, 54

Radius, 83, 90, 130
Ratio, 117; *see also* Proportion
Rectangles, 37, 38, 41, 42, 103; *see also* Golden Section; Root-five rectangles
Rhythm, 12, 14, 16
Right triangle, 73, 88, 89, 93, 98-107
Root-five rectangles, 127, 128, 129, 135, 136
Rope-stretchers, 36-47, 50, 96, 100
Rules, *see* Definitions and rules

Sakkara, 45
Samos, 62, 93, 97
Seasons, 33, 35, 42, 50
Semicircle, 53, 54, 90, 128, 129
Seshata, 46
Sextants, 54, 71
Shadow pole, 91
Shells, 19, 22
Shepherds, 30-33
Similar triangles, *see* Triangles, similar
Snowflakes, 20
Socrates, 125

Solids, regular, 106, 109-14, 119, 122
Spheres, 9, 96
Spiral, 9, 13, 17, 18, 19
Square, 17, 18, 45, 72, 103-106, 109, 132
Squaring the circle, 142
Stargazers, 33, 34, 47-56, 94
Stars, 8, 32, 51, 59, 95
Step-pyramid tomb, 45
Stone Age, 20, 21, 26, 27, 29, 34, 155
Straight angle, *see* Angles, straight
Straightedge, 7-10, 130
Sumerians, 48
Sun, 17, 24, 29, 32, 34, 50, 73, 152; eclipse of, 80
Sundial, 57, 71
Surveying, 37, 39-42; *see also* Rope-stretchers
Syene, 150, 151
Symmetry, 13, 22
Syracuse, 140, 141
Syria, 37

Taxes, 44-50
Temple of Thoth, 71
Tetrachord, 116
Tetrahedron, 110
Thales of Miletus, 62-94, 154
Theaetetus of Athens, 138, 140
Theodorus of Cyrene, 91, 140
Tigris River, 29, 36, 47, 48, 50, 59
Time, 8, 29-35, 50, 56, 155
Transit, 56
Triangles, 17, 18; equilateral, 87; inscribed, 90; isosceles, 86, 90, 109; right, 73, 84, 85, 98-107; similar, 85
Trisecting and angle, 142
Tropic of Cancer, 150

Wheel, 54

Zeno of Elea, 124
Ziggurats, 50, 67
Zodiac, 113

Titles by Jamie York Press

Making Math Meaningful®

- *Workbooks for Grades 6 – 12:*
  Student workbooks and Teacher's editions
- *A Source Book for Teaching Math in Grades 1 – 5*
- *A Source Book for Teaching Middle School Math*
- *A Source Book for Teaching High School Math*
- *Fun with Puzzles, Games and More!*

Jamie York Press

*Meaningful Math Books for Waldorf,
Public, Private, and Home Schools*
www.JamieYorkPress.com